Springer Undergraduate Mathematics Series

Springer
London
Berlin
Heidelberg
New York
Barcelona
Hong Kong
Milan
Paris
Singapore
Tokyo

Advisory Board

Other books in this series

Geoff Smith and Olga Tabachnikova

Topics in Group Theory

With 16 Figures

 Springer

Geoff C. Smith, MA, MSc, PhD
Olga M. Tabachnikova, PhD

School of Mathematical Sciences, University of Bath, Claverton Down,
Bath BA2 7AY, UK

Cover illustration elements reproduced by kind permission of:
Aptech Systems, Inc., Publishers of the GAUSS Mathematical and Statistical System, 23804 S.E. Kent-Kangley Road, Maple Valley, WA 98038, USA. Tel: (206) 432 - 7855 Fax (206) 432 - 7832 email: info@aptech.com URL: www.aptech.com
American Statistical Association: Chance Vol 8 No 1, 1995 article by KS and KW Heiner 'Tree Rings of the Northern Shawangunks' page 32 fig 2
Springer-Verlag: Mathematica in Education and Research Vol 4 Issue 3 1995 article by Roman E Maeder, Beatrice Amrhein and Oliver Gloor 'Illustrated Mathematics: Visualization of Mathematical Objects' page 9 fig 11, originally published as a CD ROM 'Illustrated Mathematics' by TELOS: ISBN 0-387-14222-3, German edition by Birkhauser: ISBN 3-7643-5100-4.
Mathematica in Education and Research Vol 4 Issue 3 1995 article by Richard J Gaylord and Kazume Nishidate 'Traffic Engineering with Cellular Automata' page 35 fig 2. Mathematica in Education and Research Vol 5 Issue 2 1996 article by Michael Trott 'The Implicitization of a Trefoil Knot' page 14.
Mathematica in Education and Research Vol 5 Issue 2 1996 article by Lee de Cola 'Coins, Trees, Bars and Bells: Simulation of the Binomial Process page 19 fig 3. Mathematica in Education and Research Vol 5 Issue 2 1996 article by Richard Gaylord and Kazume Nishidate 'Contagious Spreading' page 33 fig 1. Mathematica in Education and Research Vol 5 Issue 2 1996 article by Joe Buhler and Stan Wagon 'Secrets of the Madelung Constant' page 50 fig 1.

ISBN 1-85233-235-2 Springer-Verlag London Berlin Heidelberg

British Library Cataloguing in Publication Data
Smith, Geoff, 1953-
    Topics in group theory. - (Springer undergraduate
    mathematics series)
    1. Group theory
    I. Title II. Tabachnikova, Olga
    512.2
ISBN 1852332352

Library of Congress Cataloging-in-Publication Data
Smith, Geoff, 1953-
    Topics in group theory / Geoff Smith and Olga Tabachnikova.
        p.   cm. -- (Springer undergraduate mathematics series)
    Includes index.
    ISBN 1-85233-235-2 (alk. paper)
    1. Group Theory. I. Tabachnikova, Olga, 1967- II. Title. III. Series.
QA174.2.S65 2000
512'.2—dc21   00-037333

Typesetting: Camera ready by authors
Printed and bound at the Athenæum Press Ltd., Gateshead, Tyne & Wear
12/3830-543210 Printed on acid-free paper SPIN 10747840

To our first daughter

Preface

We very much hope that this book will be read by the interested student (and not just be parked on a shelf for occasional consultation). If you want a comprehensive reference book on Group Theory, do not buy this text. There are much better books available, some of which are mentioned below. We have a tale to tell; the absolute essentials of the theory of groups, followed by some entertainments and some more advanced material. The theory of groups is an enormous body of material which interacts with other branches of mathematics at countless frontiers. Some parts of the theory are essentially complete, but in other areas all we see are questions.

People happily read novels, so why not mathematics books? When mathematics was studied by only a few people, there was less need to try to write attractively or encouragingly since the likely readership consisted of a small group of highly motivated individuals who needed little encouragement. Even so, many talented academic writers managed to write brilliantly because they knew no other way. As higher education has opened up in economically developed countries, a much more diverse collection of people is exploring advanced mathematics and science. The challenge for authors is to produce books which engage this wider community without compromising the content.

This book is intended neither for a complete novice at Group Theory, nor for someone in the final stages of preparation to become an active researcher in the subject. Rather it is intended to support a reader engaged in a first serious group theory course, or perhaps a mathematically mature reader who is approaching group theory for the first time. It does begin at the beginning, but the beginning is not so leisurely as to suit most absolute beginners. In keeping with the spirit of the SUMS series, we have tried to make the book free-standing, and therefore usable by an independent student.

There are other texts on group theory in this world, and there are even

three in the Springer stable, so it is a serious question as to whether there is
any point in writing another. There is a chasm in Springer's coverage between
the very elementary treatment given by Wallace in *Groups, Rings and Fields*,
and the relatively sophisticated texts *An Introduction to the Theory of Groups*
by Rotman and *A Course in the Theory of Groups* by Robinson. This is the pit
into which this book is intended to fall. There are, of course, excellent books
published by other houses, and we also await Johnson's forthcoming Springer
SUMS text on symmetry, geometry and groups with keen anticipation. We are
also motivated by the surprising success of the first author's *Introductory Mathematics: Algebra and Analysis*. Its slightly irreverent tone has proved popular in
some quarters, and so we have inflicted something of the spirit of that book on
the hapless readers of this new effort. We hope that the flippancy is sufficiently
diluted by mathematics that it will not cause too much irritation.

The theory of groups is simultaneously a branch of abstract algebra, and
the study of symmetry. Since symmetry crops up in mathematics about as
often as you find trees in a wood, the jargon of group theory is part of the
workaday language of a very wide range of mathematicians and scientists. This
is not to say that all these people know a lot of group theory; some do and
some don't. The point is that if you need to talk about symmetry, you will find
group theoretic terminology very handy. A similar sentiment applies in many
contexts; you would be hard pressed to say much about space without using
points, lines, planes and angles. You would also be in deep trouble if you had
to discuss quantity (of pots, buttons, guns or knots) but you were not allowed
to use numbers.

The title of this book is in part an impudent homage to *Topics in Algebra*
(published by John Wiley). We genuflect to the spirit of Herstein for writing
such a wonderful book. We thank colleagues at the University of Bath for
comments on fragments of this book as it was being written, and pay tribute
to the many undergraduate students who engaged in careful reading and made
constructive suggestions. Sarah Anderson, Emily Harper, Damien Harwin, Paul
Lunau, Rachel Myatt, James Taylor and Petra Whittenham were particularly
assiduous in this regard. Fran Burstall and Aaron Wilson have been extremely
helpful in supporting our typesetting requirements, and Gunnar Traustason
kindly made several excellent suggestions (some of which we foolishly ignored).
We are especially indebted to the sage of the Caribbean, Dave Johnson, for
reading a draft of the text with extreme care.

We thank Andrew Swann for teaching us to drive *MetaPost* to make the
diagrams. Susan Hezlet has the credit for getting the SUMS series up and
running and David Ireland was very encouraging during the preparation of the
typescript. We also thank Karen Barker and others involved in the mathematics
section of the London office of Springer-Verlag for their continuing support.

The structure of the book is as follows. The first three chapters are intended to be relatively straightforward. Chapter 4 is conceived as an indisciplined frolic where we take the results of the first three chapters and play with them to obtain an assortment of results which we hope readers will find attractive. Chapters 5 and 6 concern more advanced topics, and by now we allow ourselves to be a little more demanding of the reader. After all, in the event that someone gets this far in, he or she should be developing a certain fluency. The first three chapters should be read in sequence. The last three chapters are intended to depend upon the first three, and Chapter 5 sometimes invokes results proved in the infinite groups section of Chapter 4. There are appendices on both the theory of fields and orderings.

A web site is associated with this book. You can download templates for making Platonic solids from there. If and when errors are detected in the text, details will be posted at the site. We are also happy to provide amplifications on any of the topics mentioned in the text upon request (provided that the scale of this enterprise does not get out of hand). We are also prepared to countenance requests to produce supplementary notes on related areas, but please look at the site first to see if your requirements are already addressed.

$$\text{http://www.bath.ac.uk/}\sim\text{masgcs/book2/}$$

Finally, we welcome our excellent daughter Marina Geoffovna to the world. She was born while this book was being written, and first heard the group axioms at the age of one minute.

GCS & OMT, Bath, 1-i-2000.

Notes for experts

(Students read this at your own peril.) Most of the material here is very standard. However, we hope that one or two points in the exposition may cause a butterfly to stir its wings. Our treatment of the alternating group includes a compelling geometric reason why no permutation can be both even and odd; a reason which involves neither determinants nor actions on polynomials. We also prove the structure theorem for finite abelian groups directly and quickly using purely group theoretic arguments (we confess to using a little number theory of the type known to Euclid). We have recourse neither to Sylow's theorem for abelian groups nor to linear algebra for this purpose. The argument hinges on the fact that the exponent of a finite abelian group is realised as the order of an element. This fact is often derived as a corollary of the structure theorem, but not here! This fast track to the proof may be of value to a lecturer who is pressed for time. There are also a couple of tricks in the appendix on finite fields which help to keep the exposition short.

We have attempted to give infinite groups considerable prominence. In the aftermath of the classification of finite simple groups, finite group theory has become less fashionable, and connections between infinite group theory and geometry are (in 2000) all the rage. Even so, the elementary theory of finite groups is very beautiful, and sheds light in so many ways on the theory of infinite groups. Finite group theory therefore continues to form an important part of undergraduate mathematics courses.

Contents

List of Figures

The Greek Alphabet

Lower Case	Upper Case	Name	Roman Equivalent
α	A	alpha	a
β	B	beta	b
γ	Γ	gamma	g (hard)
δ	Δ	delta	d
ϵ or ε	E	epsilon	e
ζ	Z	zeta	z
η	H	eta	e
θ	Θ	theta	th
ι	I	iota	i
κ	K	kappa	k
λ	Λ	lambda	l
μ	M	mu	m
ν	N	nu	n
ξ	Ξ	xi	xy (xigh in Britain, xee in the USA)
o	O	omicron	o (short)
π or ϖ	Π	pi	p
ρ	P	rho	r
σ or ς	Σ	sigma	s
τ	T	tau	t
υ	Υ	upsilon	u
ϕ or φ	Φ	phi	f
χ	X	chi	ch (as in a Scottish loch)
ψ	Ψ	psi	ps
ω	Ω	omega	o (long)

Pronunciation, particularly syllable stress, varies widely.

1
The Elements

1.1 Basic Results

In this chapter we recap basic definitions and results from the theory of groups, up to and including Lagrange's Theorem. We take this early opportunity to announce that 0 is not a natural number for the purposes of this book. Thus $0 \notin \mathbb{N}$.

The Group Axioms. A group $(G, *)$ consists of a set G and a closed binary operation $*$ such that the group axioms hold:

(i) (Associative law) For every $x, y, z \in G$ we have $(x * y) * z = x * (y * z)$.

(ii) (Identity) There is $e \in G$ such that for every $x \in G$ we have $e * x = x * e = x$.

(iii) (Inverse) For each $x \in G$ there is $x' \in G$ such that $x * x' = x' * x = e$.

To say that $*$ is a closed binary operation simply means that $x * y \in G$ whenever $x, y \in G$. It is a consequence of Axiom (i) that parentheses (brackets) serve no useful purpose in any expression of a group element as a product. For example, whenever $w, x, y, z \in G$ we have

$$(w * (x * (y * z))) = ((w * x) * (y * z)) = (((w * x) * y) * z)$$

by repeated use of the associative law. Of course, that is just an example, but the result holds in general by an induction on the number of operations in the product. Axiom (ii) asserts that a group $(G, *)$ contains an identity element. We have not insisted that e is the unique identity element, but it is. In Axiom (iii) we have not specified that the element x' be the only element of G which enjoys the given property – but again, it is. We anticipate the proof of these facts by a page or so, and call e *the* identity element of $(G, *)$ and x' *the* inverse of x. We will write the inverse of x as x^{-1} once uniqueness has been established. It would be harmless to append these uniqueness properties to the group axioms. Equally well, instead of adjoining redundant information, we could throw some away. It turns out that you can relax the above Axioms and still define a group. You can keep associativity, specify the existence of a right identity ($x * e = x$ for every $x \in G$) and an inverse on the right (given any $x \in G$ there is $x' \in G$ such that $x * x' = e$), and the full group axioms do indeed follow. Indeed, G. Higman and B.H. Neumann showed that you can get away with a single rather bizarre axiom. This result has the entertainment value (and utility) of a unicycle.

A group G is said to be *abelian* if the operation is commutative. The terminology refers to N.H. Abel. Since the word "abelian" is a proper adjective, some would argue that it must be written "Abelian", but perhaps the battle is lost. The *order* $|G|$ of $(G, *)$ is the cardinality (i.e. size) of the underlying set G. We say that the group $(G, *)$ is *finite* or *infinite* when we mean that the set G is finite or infinite. In due course we will start to refer to $(G, *)$ as just G, with the binary operation deemed implicit.

We will need to compose maps from time to time. Suppose that $\alpha : X \to Y$ and $\beta : Y \to Z$ are maps. Their composite is a map from X to Z, and we need a name for it. In many subjects, the convention is to write maps on the left of their arguments, so $\beta(\alpha(x))$ is the result of applying first α and then β to $x \in X$, and therefore it is natural to write the composition of maps as $\beta \circ \alpha$ so that $(\beta \circ \alpha)(x) = \beta(\alpha(x))$ This practice has the same merit as the typewriter keyboard; everyone is used to it. However, there is a strong (but not universal) tradition in group theory of writing maps on the right, so that the result of first applying α and then β to $x \in X$ is written $((x)\alpha)\beta$. If this is your habit, then it makes sense to write the composition of maps as $\alpha \circ \beta$ so that $(x)(\alpha \circ \beta) = ((x)\alpha)\beta$. There are two advantages to this: the eye reads the maps in the same sequence that they are composed, and the composition of permutations is easier using this convention. It is necessary to be adult and flexible about composition of maps. We are used to all sorts of notation for the application functions (maps): $\sin x$ (on the left), $x!$ (on the right), $\sqrt{x}$ (on the top and left), 2^x (to the bottom left), x^2 (to the top right) and $|x|$ (on left and right at once). The diversity of notation actually seems to help the mind to sort out the symbolism in complicated expressions. Our default position will be that maps are written on the right, but we shall write maps on the left (or anywhere we please) when there is some reason to do so. For example, linear algebra (matrices, vectors) is a subject used by virtually all practitioners of science and pseudoscience. Many people have the habit of writing their linear transformations (matrices) on the left of their vectors (column n-tuples), and this forces composition from right to left; ST means first apply the linear transformation (matrix) T, and then the linear transformation (matrix) S. There is no point in fighting this, and we happily go with the crowd.

The results of the next lemma are at the heart of the theory of groups; these facts will be used so frequently that normally we shall make no reference to this lemma.

Lemma 1.1 (Cancellation)

Suppose that $(G, *)$ is a group.

(i) If $x, y, z \in G$ and $x * y = x * z$, then $y = z$.

(ii) If $x, y, z \in G$ and $y * x = z * x$, then $y = z$.

(iii) If $x, y \in G$, then there is $z \in G$ such that $y = x * z$.

(iv) If $x, y \in G$, then there is $w \in G$ such that $y = w * x$.

(v) If $x, y \in G$ and $x * y = x$, then $y = e$.

(vi) If $x, y \in G$ and $y * x = x$, then $y = e$.

If the following proof seems too abstract for your tastes, have no fear. We will discuss a more direct approach after Corollary 1.3.

Proof

Recall that a map $f : X \to X$ is a bijection if and only if there is a map $g : X \to X$ such that $f \circ g = g \circ f = \mathrm{Id}_X$, the identity map on X.

Now for part (i). Fix $x \in G$ and consider the map $f : G \to G$ defined by $f : y \mapsto x * y$ for every $y \in G$, and the map $g : G \to G$ defined by $g : y \mapsto x' * y$ for every $y \in G$. For any $y \in G$ we have

$$(y)(g \circ f) = ((y)g)f = (x' * y)f = x * (x' * y) = (x * x') * y = e * y = y.$$

The equalities are justified as follows: definition of $g \circ f$, definition of g, definition of f, Axiom (i), definition of x' and definition of e. Thus $g \circ f = \mathrm{Id}_G$. The proof that $f \circ g = \mathrm{Id}_G$ is obtained by exchanging the roles of x and x'.

We conclude that f is bijective. The facts that f is injective and surjective translate to give (i) and (iii). One can prove (ii) and (iv) in a similar way. The results (v) and (vi) are just special cases of (i) and (ii) applied to $x * y = x * e$ and $y * x = e * x$.

$\square$

Corollary 1.2

If e' is a rival identity element to e, then $e = e * e' = e'$. If x'' is a rival inverse to x' then $x * x' = e = x * x''$ so by part (i) we deduce that $x' = x''$.

It is now completely legitimate to talk about *the* identity element of G and *the* inverse of an element $x \in G$, and henceforth we allow ourselves the notation x^{-1}.

Corollary 1.3

Suppose that $x * y = e$. We also have $x * x^{-1} = e$ and Lemma 1.1 part (i) applies so $y = x^{-1}$. We conclude that $y * x = e$. Similarly, if $y * x = e$, then $x * y = e$ and once again $y = x^{-1}$.

A reader who is unnerved by the use of maps in proving Lemma 1.1 can take comfort in the fact that the results can all be established using "bare hands". For example, to prove part (i) one can reason as follows. Suppose that $x * y = x * z$. By Axiom (iii) there exists an inverse x' to x. Thus $x' * (x * y) = x' * (x * z)$, and by Axiom (i) we have that $(x' * x) * y = (x' * x) * z$ so by Axiom (ii) we have $e * y = e * z$ and therefore $y = z$. The reader should supply similar proofs of the other parts of Lemma 1.1, and in fact this is the content of the first exercise. Solutions to all exercises (at least in kit form) may be found at the back of this book.

EXERCISES

1.1 Prove parts (ii)–(vi) of Lemma 1.1 without the use of maps.

1.2 Suppose that G is a group and that $|G| = 2$. Show that $g * g = e$ for every $g \in G$.

1.3 Suppose that G is a group and that $|G| = 3$. Show that $g * g * g = e$ for every $g \in G$.

1.4 Consider L the set of 26 lower case Roman letters $\{a, b, c, \ldots, x, y, z\}$. Define a "multiplication" $*$ on L by $\alpha * \beta = \alpha$ for every $\alpha, \beta \in L$. Which fragments of the group axioms are valid for this structure?

1.5 Suppose that G is a group with binary operation $*$. Define a new operation $\Box$ on the set G by $x \Box y = y * x$. Show that G equipped with $\Box$ is a group.

1.6 Justify the assertion made in the text that Axioms (ii) and (iii) may be relaxed so that you only need specify the existence of a right identity and right inverses, and yet still define a group. The reader may wish to contemplate the expression $x' * x * x' * (x')'$.

We have already remarked that we can dispense with brackets. We can and will also dispense with the symbol $*$, and will usually write xy instead of $x * y$, and G instead of $(G, *)$. We will also casually refer to the group operation as "multiplication". If we really need to emphasize the presence of a product, a small central dot $x \cdot y$ will be inserted. We allow ourselves to write 1_G or just 1 for the (unique) identity element of G. In the event that the operation is commutative, we may write the operation as $+$, and call it "addition". It then makes sense to write the identity element as 0_G or 0, and the inverse of $x \in G$ as $-x$. Of course we could use any notation we please for the group operation, the identity element and inverses, but it is sensible to make choices which complement our previous experience of how symbols behave. For example, there is nothing to stop you using $\div$ for the group operation, the copyright symbol $\copyright$ for the identity element and $\$x$ for the inverse of x. However, readers would probably find such notation unsettling and ridiculous.

Lemma 1.4

Suppose that G is a group.

 (i) The map $\mathrm{inv} \colon G \to G$ defined by $g \mapsto g^{-1} \ \forall g \in G$ is a bijection.

 (ii) If $x, y \in G$, then $(xy)^{-1} = y^{-1}x^{-1}$.

Proof

 (i) This follows from Corollary 1.2 since $\mathrm{inv} \circ \mathrm{inv} = \mathrm{Id}_G$.

 (ii) Observe that $xyy^{-1}x^{-1} = y^{-1}x^{-1}xy = 1$. It follows that $(xy)^{-1} = y^{-1}x^{-1}$ by uniqueness of inverses.

$\square$

Corollary 1.5

If $x_1, x_2, \ldots, x_n \in G$, then $(x_1 x_2 \cdots x_n)^{-1} = x_n^{-1} x_{n-1}^{-1} \cdots x_1^{-1}$ by induction on n.

Remark 1.6

Suppose that G is a group and that $x \in G$. We allow the exponential notation x^n for $n \in \mathbb{Z}$ where $x^1 = x$, and for $n > 1$ we make the inductive definition $x^n = x^{n-1}x$. We put $x^0 = 1$, and for negative integers n we inductively define $x^n = x^{-1}x^{n+1}$.

In order to deduce the properties of this symbolism, we must resist the temptation to think that y^n means y *multiplied by itself n times*. That thought is dangerous since n can be 0, 1 or negative. We issue a warning about the proof of Lemma 1.7. Because of our familiarity with the results stated in this lemma, we have to be extremely careful with the proof to ensure that we are not inadvertently using some property of the exponential notation which is yet to be established. There is the relaxed alternative of regarding the result as obvious and skipping the proof. Your conscience must decide.

Lemma 1.7

Suppose that x is an element of a group G.

(i) $x^n \cdot x^{-n} = 1$ for every $n \in \mathbb{Z}$

(ii) $x^n \cdot x = x \cdot x^n$ for every positive integer n.

(iii) $(x^{-1})^n = (x^n)^{-1} = x^{-n}$ for all integers n.

Proof

Part (i) holds when $n = 0$, and when $n > 0$ we have

$$x^n \cdot x^{-n} = x^{n-1}x \cdot x^{-1}x^{-(n-1)} = x^{n-1} \cdot x^{-(n-1)} = 1.$$

The last equality is by induction on n. Now also $x^{-n} \cdot x^n = 1$ for $n \geq 0$ by Corollary 1.3. We conclude that $x^n \cdot x^{-n} = 1$ for every $n \in \mathbb{Z}$.

Next we address part (ii). When $n = 1$ there is nothing to prove so we may assume that $n > 1$ and induct on n. Now $x^n \cdot x = (x^{n-1} \cdot x) \cdot x = (x \cdot x^{n-1}) \cdot x$. The equalities are justified by definition and by induction respectively. Now associativity and the definition of x^n yield that $x^n \cdot x = x \cdot (x^{n-1} \cdot x) = x \cdot x^n$ by inductive hypthesis, and this part is done.

Finally we tackle part (iii). We show that for every $n \in \mathbb{Z}$ the quantities x^n and $(x^{-1})^n$ are mutually inverse. We note that the result holds when $n = 0$ and induct on $n \geq 0$ as usual. We may assume that $n > 0$. Now

$$x^n \cdot (x^{-1})^n = (x^{n-1}x) \cdot (x^{-1})^{n-1}x^{-1}$$

by definition of the exponential notation. We deploy part (ii) and induction to obtain that

$$x^n \cdot (x^{-1})^n = (xx^{n-1}) \cdot (x^{-1})^{n-1} x^{-1} = x \cdot (x^{n-1} \cdot (x^{-1})^{n-1}) \cdot x^{-1} = x \cdot 1 \cdot x^{-1} = 1.$$

For non-negative n we are done. Now take the equation $x^n \cdot (x^{-1})^n = 1$ when $n \geq 0$ and invert it using part (i). We obtain that. $(x^{-1})^{-n} \cdot x^{-n} = 1$ for $n \geq 0$. Replace x by x^{-1} and $-n$ by m to obtain that $x^m \cdot (x^{-1})^m = 1$ whenever $m \leq 0$. We conclude that $x^n \cdot (x^{-1})^n$ for every $n \in \mathbb{Z}$. Thanks to this latest result and part (i) we know that for every integer n, the elements x^{-n} and $(x^{-1})^n$ are both inverse to x^n. Uniqueness of inverses forces $x^{-n} = (x^n)^{-1} = (x^{-1})^n$.

$\square$

More generally $x^n x^m = x^{n+m}$ and $(x^m)^n = x^{mn} = (x^n)^m$ for all $m, n \in \mathbb{Z}$. If you have a sunny disposition, you may regard these results as obvious, but the more critical reader deserves a proper argument.

Proposition 1.8

Suppose that $m, n \in \mathbb{Z}$ and that x is an element of a group G.

(i) $x^m x^n = x^{m+n}$ and

(ii) $(x^m)^n = x^{mn} = (x^n)^m$.

Proof

(i) There is little difficulty if $m, n \geq 0$, but even in this case we must prove the result. We use induction on n. If $n = 0$ the result is true, so the base case is done. Now for the induction step, assuming that $n > 0$. Now $x^m x^n = x^m (x^{n-1} x)$ (using the definition of x^n) so the associative law yields $x^m x^n = (x^m x^{n-1}) x = x^{m+n-1} x$ by inductive hypothesis, and so $x^m x^n = x^{m+n}$.

Now we address the remaining cases, when at least one of m and n is negative. Consider the equation $x^{-n} x^{-m} = x^{-(m+n)}$. If $m, n < 0$ then $-m, -n \geq 0$ and so this equation is correct because we have already dealt with this situation. Now $x^m x^n = (x^{-n} x^{-m})^{-1}$ by Lemma 1.7 so $x^m x^n = (x^{-(m+n)})^{-1} = x^{m+n}$ as required.

There are another four cases to consider. The following table displays the key equation in each case.

m	n	$m+n$	Equation	
≥ 0	< 0	≥ 0	$x^{m+n}x^{-n}$	$= \quad x^m$
≥ 0	< 0	< 0	$x^{-(m+n)}x^m$	$= \quad x^{-n}$
< 0	≥ 0	≥ 0	$x^{-m}x^{m+n}$	$= \quad x^n$
< 0	≥ 0	< 0	$x^n x^{-(m+n)}$	$= \quad x^{-m}$

Each of the four equations under discussion is just part (i) of the proposition in a context where exponents are not negative. We finish the proof of part (i) by manipulating each of the four equations in an appropriate way: postmultiply the first equation by x^n, premultiply the second equation by x^{m+n} and postmultiply it by x^n, premultiply the third equation by x^m, and finally take the fourth equation, premultiply it by x^m and postmultiply it by x^{m+n}. In every case we appeal to Lemma 1.7

(ii) If either of m or n is 0 the result holds. If $m, n > 0$ the result can be demonstrated by induction on n. If $m, n < 0$, then $(x^{-m})^{(-n)} = x^{mn}$ by the case of positive exponents, However

$$(x^m)^n = \left(\left((x^{-1})^{-1} \right)^m \right)^n = (x^{-m})^{(-n)}$$

thanks to repeated application of part (iii) of Lemma (1.7).

The cases $n < 0 < m$ and $m < 0 < n$ are dealt with similarly.

$\square$

Corollary 1.9

If G is a group and $x \in G$, then $x^m x^n = x^{m+n} = x^{n+m} = x^n x^m$ for all $m, n \in \mathbb{Z}$. Thus any powers of the same element commute.

We now introduce the notion of a *subgroup*. Suppose that G is a group and $H \subseteq G$. We say that H is a subgroup of G when the group operation of G when restricted to elements of H renders H a group. The group operation on G is a map $G \times G \to G$. Since $H \times H \subset G \times G$ we have have an induced map $H \times H \to G$. Inversion can also be viewed as a map $G \to G$ yielding an induced map $H \to G$. The subgroup property amounts to saying that H is non-empty and the images of each of the two induced maps are contained in H. Notice that associativity is inherited from G, so never causes a problem. We now make this formal.

Definition 1.10

Suppose that G is a group. A subset H of G is a *subgroup* of G if and only if the following three conditions are satisfied:

(a) $1_G \in H$.

(b) If $x, y \in H$, then $xy \in H$.

(c) If $x \in H$, then $x^{-1} \in H$.

A subgroup H is *proper* if $H \neq G$.

If H is a subgroup of G we write $H \leq G$ or $G \geq H$. If in addition $H \neq G$ we may write $H < G$ or $G > H$. Notice that G and $\{1\}$ are subgroups of G. The latter is called the trivial subgroup, and is often written 1 (deliberately abusing notation). When writing an abelian group additively, it makes sense to abuse notation in a consistent manner, so we write 0 rather than $\{0\}$ for the trivial subgroup in this context.

Suppose that S is a non-empty subset of G which happens to be a group when we restrict the given group operation on G to S. At first sight it might seem possible that a multiplicatively closed subset S of G could form a group in its own right, but have an identity element 1_S different from the identity element of G. Notice that $1_S 1_S = 1_S = 1_S 1_G$ so Lemma 1.1(i) yields $1_S = 1_G$. Thus $1_G \in S$ and our concern is shown to have no foundation. However, we have only escaped by the skin of our teeth (this is not a book on anatomy) because in other algebraic contexts, this sort of thing can happen.

Proposition 1.11

Suppose that G is a group and that $H \subseteq G$. It follows that H is a subgroup of G if and only if both

(α) $H \neq \emptyset$ and

(β) if $x, y \in H$, then $xy^{-1} \in H$.

Proof

The Greek letter labels in this proposition are there to help avoid confusion with the Roman labels of Definition 1.10. It is clear that when H is a subgroup of G, conditions (α) and (β) are both satisfied.

Next we assume that (α) and (β) both hold. We must show that conditions (a), (b) and (c) of Definition 1.10 are all satisfied. By (α) we can choose $h \in H$. Put $x = y = h$ in (β) to deduce that $1 = hh^{-1} \in H$. Now put $x = 1$ and $y = h$

in (β) to deduce that $1h^{-1} = h^{-1} \in H$. We have established that conditions (a) and (c) both hold. Next we tackle condition (b). Suppose that $x, y \in H$, so $y^{-1} \in H$ since we have shown that H is closed under inversion. Now apply condition (β) so $x(y^{-1})^{-1} = xy \in H$ and we are done.

$\square$

Every group G has subgroups 1 and G, and these subgroups are different unless $G = 1$. We learn nothing about G from these extreme cases. Subgroups are very natural objects to reason about, so we might hope that the intersection and union of a pair of subgroups should be subgroups. However, in general only one of these hopes is realised.

Proposition 1.12

Suppose that G is a group and G_λ is a subgroup of G for each $\lambda \in \Lambda$. It follows that $\bigcap_{\lambda \in \Lambda} G_\lambda$ is a subgroup of G.

N.B. The set Λ is just an indexing set, a set of labels for the subgroups in question. In the case that Λ has size 2, this proposition simply asserts that the interection of two subgroups of G is a subgroup of G. The generalization we have given is valid even when Λ is uncountably infinite.

Proof

First we observe that $1 \in G_\lambda$ for all $\lambda \in \Lambda$, so $1 \in M = \bigcap_{\lambda \in \Lambda} G_\lambda \neq \emptyset$. Now suppose that $x, y \in M$. Thus $x, y \in G_\lambda$ for every $\lambda \in \Lambda$, so $xy^{-1} \in G_\lambda$ for every $\lambda \in \Lambda$. It follows that $xy^{-1} \in M$ whenever $x, y \in M$. By Proposition 1.11 we are done.

$\square$

EXERCISES

1.7 Give a proof of part (i) of Lemma 1.4 which does not use the characterization of bijections in terms of maps.

1.8 Suppose that G is an abelian group. Prove that each of the following subsets of G is a subgroup.

(a) Fix a natural number n, and let $A = \{x \mid x \in G, x^n = 1\}$.

(b) $B = \{y \mid \exists z \in G \text{ with } y = z^n\}$.

(c) $C = \{w \mid w^m = 1 \text{ for some } m > 0\}$.

1.9 Suppose that G is a group containing a subgroup H. Choose any $g \in G$. Show that $D = \{g^{-1}hg \mid h \in H\}$ is a subgroup of G.

1.10 Suppose that G is a group and H_i $(i \in I)$ is a non-empty family of subgroups of G (I is acting as an indexing set). Suppose furthermore that whenever $j, k \in I$, either $H_j \leq H_k$ or $H_k \leq H_j$. Show that $\cup_{i \in I} H_i$ is a subgroup of G.

Since a group consists of a set equipped with a "multiplicative" structure, you might think it sensible to try to define groups using multiplication tables.

Example 1.13

Consider the following multiplication table:

$$
\begin{array}{c||cccccc}
 & e & \alpha & \beta & \gamma & \delta & \varphi \\
\hline\hline
e & e & \alpha & \beta & \gamma & \delta & \varphi \\
\alpha & \alpha & e & \varphi & \delta & \gamma & \beta \\
\beta & \beta & \delta & e & \varphi & \alpha & \gamma \\
\gamma & \gamma & \varphi & \delta & e & \beta & \alpha \\
\delta & \delta & \beta & \gamma & \alpha & \varphi & e \\
\varphi & \varphi & \gamma & \alpha & \beta & e & \delta \\
\end{array}
\tag{1.1}
$$

This table defines a binary operation on $S = \{e, \alpha, \beta, \gamma, \delta, \varphi\}$ where each entry denotes the product of its row label with its column label, in that order. Thus, for example, $\alpha\delta = \gamma$ whereas $\delta\alpha = \beta$. Note also that we want S to be a set of size 6, so the symbol e and the lower case Greek letters denote distinct elements of the set.

Is S a group when equipped with this closed binary operation? The symbol e is clearly the identity from Table (1.1), and inverses exist by inspection. However, the truth of the associative law may not at first be apparent. After all, there are 216 instances of the equation $(xy)z = x(yz)$ where x, y and z range over the elements of S. A sad neglected individual might wish to fill an empty hour by checking each one. However, we suggest a better plan. Instead of trying to define a group from the table, we find a group which has this multiplication table. By doing this we ensure that the associative law holds. Note that while a table is a perfectly good way to record the multiplicative structure of a group, we will always be confronted with this associativity problem if we try to use a table to define a group.

Let $\Omega = \{1, 2, 3\}$, and define six bijections from Ω to Ω as follows.

$$i : \begin{cases} 1 \to 1 \\ 2 \to 2 \\ 3 \to 3 \end{cases} \qquad a : \begin{cases} 1 \to 2 \\ 2 \to 1 \\ 3 \to 3 \end{cases} \qquad b : \begin{cases} 1 \to 1 \\ 2 \to 3 \\ 3 \to 2 \end{cases}$$

$$c : \begin{cases} 1 \to 3 \\ 2 \to 2 \\ 3 \to 1 \end{cases} \qquad d : \begin{cases} 1 \to 2 \\ 2 \to 3 \\ 3 \to 1 \end{cases} \qquad f : \begin{cases} 1 \to 3 \\ 2 \to 1 \\ 3 \to 2 \end{cases} .$$

Let $\Gamma = \{i, a, b, c, d, f\}$, so Γ consists of the six possible bijections from Ω to Ω. Note that $a \circ b$ means first apply a, then apply b since we are working with permutations.

Thus $(1)(a \circ b) = ((1)a)b = (2)b = 3$. If we rename the six elements of Γ in the obvious way: i is e, a is α, b is β, c is γ, d is δ and f is φ you can easily verify that the multiplication table (1.1) is correct.

The set Γ has size $3! = 6$ and is closed under composition of maps and under inversion. Moreover, the identity map i acts as a two-sided identity, and composition of maps is always associative. Thus Γ forms a group using map composition as the operation. Since the group S which we tried to define using a multiplication table is really just a renaming of the elements of Γ, it follows that the multiplicative structure we defined on S is associative, and there is no need to check 216 conditions after all. If Ω is any set, then we can construct a group in this way. If $|\Omega| = n$ is finite, the group of all bijections from Ω to Ω will have size $n!$, and in fact we can even let Ω be empty, in which case the group of bijections is trivial[1]. We will have more to say about this construction in Section 1.2.

A multiplication table of a group is often called its *Cayley Table*. Cayley tables have certain limitations.

(i) It may be tricky to tell from the table that the operation is associative, though Axioms (ii) and (iii) are easier to verify.

(ii) If the group is finite but large, the table may be unwieldy. For example, perhaps there are 100 students majoring in (reading) mathematics at your institution. At the request of other students and the public health authorities, they are housed separately from the rest of the student body. In order to allow the mathematics students to concentrate fully on their discipline, the campus accommodation office attempts to disrupt friendships. Therefore, every Friday afternoon, 100 postcards arrive in the mathematics accommodation pigeon-holes instructing the occupant of each room i to move

---

[1] This is why $0! = 1$

to room $f(i)$ on Sunday evening. A highly paid team of mapping managers makes up a function f each week, overseen by the injectivity inspectorate, the surjectivity surveillance squad and a quality assurance team from the bijectivity bureau. This apparatus is ostensibly there to make sure that each f is a bijection, but since the procedure only involves checking that the function has been sent out on postcards of the correct size, this does not always work.

The set of all 100! possible bijections forms a group under composition. If you wish to describe this group using a multiplication table, in addition to supplying the row and column labels you will need fill in $(100!)^2 =$

87097824890894800794165901619444858655697206439408401342
15932536243379996346583325877967096332754920644690380762
21960747636428941143592019057396067750788139460748990533
17297580134329929871847646073758894343134833829668015151
56280854162691766195737493173453603519594496000000000000
000000000000000000000000000000000000

entries.

(iii) If the group G is infinite, then one cannot read or write the multiplication table in finite time. Worse yet, if G is uncountable, then you cannot list the elements of G as labels for columns. The relationship between a group multiplication law and its Cayley table is very similar to the relationship between a function and its graph. Small children learn tables for multiplying some whole numbers but it would not really improve your understanding of the function sine : $\mathbb{R} \to \mathbb{R}$ if you tried to memorize the value of $\sin x$ for every $x \in \mathbb{R}$. Such information does specify the sine function of course, but there is an uncountable amount of data, and it is not presented in a very useful form. When people do mathematics with the function sine, they limp along by knowing (a) the general shape of the graph of sine, (b) some handy relationships between sine and other functions (for example $\sin(x + y) = \sin x \cos y + \cos x \sin y$) and (c) that if given $r \in \mathbb{R}$ to some specified degree of accuracy, they can work out $\sin r$ to a computable degree of accuracy. In the case of groups, as soon as there are more than about 10 or 12 elements of a group, the multiplication table becomes a little overwhelming for most humans. However, we can understand groups in other ways.

1.2 Where Do Groups Come From?

Groups arise in mathematical nature as follows. Let Ω be a set, and let H be a collection of bijections from Ω to Ω. We let $\circ$ denote composition of bijections. It may so happen that $(H, \circ)$ satisfies the group axioms. For example, the so-called symmetric group on Ω is $(\text{Sym } \Omega, \circ)$ where Sym Ω is the set of all bijections from Ω to Ω. Notice that if $f \in \text{Sym } \Omega$, then f^{-1} has two meanings; the function which is inverse to f, and the group element which is inverse to f. Happily for all concerned, these two things coincide, so even the most sensitive reader should feel able to cope with this ambiguity.

When $\Omega = \{1, 2, \ldots n\}$, the set of the first n natural numbers, the group Sym Ω is often written S_n. We say that this is the *symmetric group of degree* n, and its order is $n!$ (pronounced n factorial or sometimes factorial n). We will discuss degree more carefully later (see Definition 3.1), but informally we say that S_n has degree n because it consists of bijections from a set of size n to itself. Notice that S_1 is the trivial group, and that S_2 is an abelian group. However, S_3 is not abelian, and it is a group that we have already investigated in Example 1.13. We can think of S_n as the subgroup of S_{n+1} consisting of those bijections which fix $n + 1$. It follows that S_m is non-abelian whenever $m \geq 3$.

Suppose that the set Ω carries some extra structure. We focus upon those bijections from Ω to Ω which preserve structure and which, in addition, have inverses which also preserve structure. Such a collection of maps will form a group under composition. Recall that composition of maps is associative, which is one less thing to worry about. We have deliberately not defined the word *structure*.

We examine some explicit examples. The first one uses the notion of an ordered space. Appendix B includes a quick summary of the theory of relations and orderings.

Example 1.14

A *totally ordered space* $(T, <)$ is a set T which comes equipped with a relation $<$ satisfying the following axioms.

(i) For all $s, t \in T$, exactly one of the following three statements is true:
   (a) $s < t$,   (b) $t < s$,   (c) $s = t$.

(ii) If $s, t, u \in T$ are such that both $s < t$ and $t < u$, then $s < u$.

An example of a totally ordered space is $(\mathbb{Q}, <)$, the rational numbers equipped with the usual ordering. Here we have discarded the additive and multiplicative structures of $\mathbb{Q}$, and are just thinking of $\mathbb{Q}$ as an ordered set. Let A be the set of bijections from $\mathbb{Q}$ to $\mathbb{Q}$ which preserve the ordering, so $\gamma \in A$ if and only if $\gamma : \mathbb{Q} \to \mathbb{Q}$ is a bijection such that if $q_1 < q_2$, then $(q_1)\gamma < (q_2)\gamma$. Examples of elements of A include the map $q \mapsto q + 1$ (with inverse $q \mapsto q - 1$), and the map $q \mapsto 2q$ (with inverse $q \mapsto q/2$).

We first show that the composition of order-preserving maps preserves the order. Suppose that $\mu, \nu \in A$, and that $r, s \in \mathbb{Q}$ with $r < s$, then $(r)\mu < (s)\mu$, so $((r)\mu)\nu < ((s)\mu)\nu$. This means that $(r)(\mu \circ \nu) < (s)(\mu \circ \nu)$ as required.

We next show that the inverse of an order-preserving bijection preserves the order. Suppose that $\mu \in A$, and that $r, s \in \mathbb{Q}$ with $r < s$. Now μ is a bijection so $r = (u)\mu$ and $s = (v)\mu$ for unique and distinct $u, v \in \mathbb{Q}$. It cannot be that $v < u$, for then it would follow that $(v)\mu < (u)\mu$, or rather $s < r$, which is absurd. Thus $u < v$. But $u = (r)\mu^{-1}$ and $v = (s)\mu^{-1}$ so $(r)\mu^{-1} < (s)\mu^{-1}$ and we have shown that μ^{-1} preserves order, so the order-preserving bijections from $\mathbb{Q}$ to $\mathbb{Q}$ form a group A under composition.

Notice that in this case it so happens that every structure-preserving bijection has a structure preserving inverse. When doing algebra, this is what you should expect. However, if the structure in question is sufficiently rich, then this may not happen. *If you don't know the meaning of the following sentence, please ignore it.* A continuous bijection between topological spaces need not have a continuous inverse.

EXERCISES

1.11 Show by example that there is a non-bijective map $\gamma : \mathbb{Q} \to \mathbb{Q}$ which has the property that whenever $q_1, q_2 \in \mathbb{Q}$ and $q_1 < q_2$, then $(q_1)\gamma < (q_2)\gamma$.

1.12 Demonstrate that the group A of Example 1.14 has the following interesting properties.

(a) A is group.

(b) If $\gamma \in A$ and $\gamma \neq 1_A$, then as i runs through the integers, the elements γ^i are all distinct.

Example 1.15

To follow this example, it is necessary to understand the basics of linear algebra. Let V be a vector space over a field k. If the linear map $S : V \to V$ happens to be bijective, then we say that S is a non-singular linear transformation of V. Let B be the collection of all non-singular linear transformations of V. We now show that B is a group under the operation "composition of functions". Recall that $S : V \to V$ is a linear transformation if and only if for all $\lambda, \mu \in k$ and for all $u, v \in V$, we have

$$S(\lambda u + \mu v) = \lambda \cdot (S(u)) + \mu \cdot (S(v)).$$

Note that we are writing maps on the left and composing maps from right to left because this is the polite thing to do in linear algebra land.

(a) The composition of bijective maps is bijective. It remains to verify that if X, Y are linear transformations, then so too is XY. Suppose that $\lambda, \mu \in k$ and $u, v \in V$. Now

$$(XY)(\lambda u + \mu v) = X(Y(\lambda u + \mu v)) = X \left(\lambda \left(Y(u) \right) + \mu \left(Y(v) \right) \right)$$

$$= \lambda \left(X \left(Y(u) \right) \right) + \mu \left(X \left(Y(v) \right) \right) = \lambda(XY)(u) + \mu(XY)(v).$$

(b) The identity map Id_V is in $\mathrm{GL}(V)$.

(c) The inverse of a bijection is a bijection, and it remains to check that the inverse map to a non-singular linear transformation S is a linear transformation. Again we suppose that $\lambda, \mu \in k$ and $u, v \in V$. Now, S is bijective, so there are unique $w, x \in V$ such that $u = S(w)$ and $v = S(x)$. We now have

$$S^{-1}(\lambda u + \mu v) = S^{-1}(\lambda S(w) + \mu S(x)) = S^{-1} \left(S(\lambda w + \mu x) \right)$$

$$= \lambda w + \mu x = \lambda S^{-1}(u) + \mu S^{-1}(v),$$

and so S^{-1} is a linear transformation.

We conclude that B is a group. The usual name for this group is the *general linear* group of V, written $\mathrm{GL}(V)$. The elements are the structure-preserving bijections from V to V, but this time the structure in question is the vector space of which V is the underlying set. The word "general" is used because we are embracing all of the structure-preserving bijections from V to V, rather than just some of them.

Suppose that V happens to be finite dimensional. We intrude on the privacy of V by picking a basis $v_1, v_2, \ldots, v_n$ and then associating to each $S \in \mathrm{GL}(V)$ an $n \times n$ matrix $A = (a_{ij})$ via the system of equations

$$S(v_j) = \sum_{i=1}^{n} a_{ij} v_i \text{ for } j = 1, 2, \ldots, n. \tag{1.2}$$

This sets up a bijection between the group $\mathrm{GL}(V)$ and $\mathrm{GL}_n(k)$, the multiplicative group of all $n \times n$ invertible matrices with entries in k. Moreover, this bijection respects the group structures in the sense that if you fix the basis, and contemplate $X, Y \in \mathrm{GL}(V)$ with associated matrices $\overline{X}$ and $\overline{Y}$, then the matrix of XY is $\overline{XY} = \overline{X}\,\overline{Y}$. In this way, matrices allow explicit calculation in $\mathrm{GL}_n(k)$, a group which is simply a copy of $\mathrm{GL}(V)$. There is a price to pay of course. Matrices can be large and unwieldy objects. Moreover, we have disturbed V by selecting a basis. If we worked with a rival basis $u_1, \ldots, u_n$, then the story would unfold in a similar way, except that we would not necessarily get the same bijection between $\mathrm{GL}(V)$ and $\mathrm{GL}_n(k)$. It is useful to know the relationship between these bijections. Recall that Eq.(1.2) tells us how the matrix (a_{ij}) of S is obtained using the basis $v_1, \ldots, v_n$.

Choose a rival basis $u_1, \ldots, u_n$ and suppose that the matrix of S with respect to this basis is $B = (b_{ij})$, so the entries of this matrix are obtained from the equations

$$S(u_j) = \sum_{i=1}^{n} b_{ij} u_i \text{ for } j = 1, 2, \ldots, n. \tag{1.3}$$

The relationship between the matrices A and B is as follows. There are $P, Q \in \mathrm{GL}_n(k)$ such that $QAP = B$ and $PBQ = A$. The matrices $P = (p_{ij})$ and $Q = (q_{ij})$ are mutually inverse, and arise from the systems of equations

$$u_j = \sum_{i} p_{ij} v_i \text{ for } j = 1, 2, \ldots, n \tag{1.4}$$

and

$$v_j = \sum_{i} q_{ij} u_i \text{ for } j = 1, 2, \ldots, n. \tag{1.5}$$

It follows that $P = Q^{-1}$ and $Q = P^{-1}$, so there are various ways to express the equivalent matrix equations $QAP = B$ and $PBQ = A$. We may write $QAQ^{-1} = B$, $P^{-1}AP = B$, $Q^{-1}BQ = A$ and $PBP^{-1} = A$. None of these equations has any preferred status. This causes havoc in linear algebra since different authors choose one equation rather than the others to capture the relationship between A and B. We give you the lot, and ask you to take your pick. A word in your ear however – we are currently writing maps on the left, so the mathematics may be a little smoother if you use $QAQ^{-1} = B$ or $PBP^{-1} = A$. We will expand on this point later in Remark 3.27.

Notice that the determinant of the matrix of $S \in \mathrm{GL}(V)$ is independent of the choice of basis because

$$\det (CD) = \det(C) \cdot \det(D) = \det(D) \cdot \det(C) = \det(DC)$$

for all $n \times n$ matrices C and D, and therefore

$$\det A = \det (PBP^{-1}) = \det (P^{-1}PB) = \det B.$$

Thus we may talk about the determinant of $S \in \mathrm{GL}(V)$. The elements of $\mathrm{GL}(V)$ of determinant 1 form a subgroup denoted $\mathrm{SL}(V)$, the special linear group of V. There is a corresponding matrix group $\mathrm{SL}_n(k)$.

Example 1.16

Let $\Omega \subset \mathbb{R}^3$ be a unit cube (so each of the 12 edges of Ω has length 1) in 3-dimensional Euclidean space. For exactness, we shall assume that $\Omega = [0, 1]^3$. An *isometry* (of the cube) is a map from Ω to Ω which preserves distances. It is an exercise to show that isometries are necessarily bijections in this context. As the reader might expect, the composition of isometries is an isometry, and the inverse of an isometry is an isometry. These facts should also be checked. We have a group C of all the isometries. Since the corners are the only points of the cube with the property that they have distance $\sqrt{3}$ from another point of the cube, these isometries must send corners to corners, and the reader is also invited to show that an isometry is determined by what it does to the corners (i.e. if two isometries do the same thing to the corners, then they must in fact be the same isometry).

These isometries come in two types. There are 24 isometries that can be accomplished by a rigid motion of the cube, and 24 which need a reflection in addition to the rigid motion. We use the term *rigid motion* in a semi-formal way; we mean a bijection θ from the cube to itself which has the following property:

The cube can be moved continuously through 3-dimensional Euclidean space in such a way that the distance between each pair of points in the cube does not change during the movement, and when the movement stops, each point p of the cube is now in position $(p)\theta$.

Reflection in a plane does preserve distances, but cannot be accomplished by a rigid motion because if the surface of the cube is decorated in an irregular way, then the orientation (the "handedness") of the artwork changes under reflection, but is preserved under a rigid motion. If you find yourself wondering if you or your image in a mirror is actually real, you have the option of breaking your symmetry just to be sure. A tatoo or some minor body piercing will suffice.

We concentrate on the subgroup D of isometries which consists of the rigid motions. The reason that D has order 24 is that a rigid motion can put any face underneath (6 choices), and then rotate the cube into any one of four positions. Thus D is a finite group. We will describe its elements, but will refrain from giving a multiplication table.

The cube has 4 long diagonals, and rotation through $2\pi/3$ in either direction about one of these axes is an element of D. This yields 8 elements.[2] There are 3 axes joining centres of faces to centres of opposite faces. There are 6 elements of D corresponding to rotation through $\pm\pi/2$ about these axes, and another 3

---

[2] This has nothing to do with group theory, but Prof D.L. Johnson of Jamaica has pointed out to me that it is interesting to calculate the volume of revolution of a cube spinning round a long diagonal. Viewed as a fraction of the volume of the circumscribing ball, the result sends a chill down one's spine.

elements corresponding to rotation through π about the same axes. There are 6 axes joining the midpoint of an edge to the midpoint of the opposite edge. These axes give rise to 6 elements of D which are rotations through π about these axes. Finally there is the identity element of D. The reader should quickly check that the 24 elements we have described are genuinely different, and then he or she will know that we do have a complete list of the elements of D. In fact the group D is a disguised copy of S_4, the symmetric group of degree 4. We will explain why this is true in Example 3.25. There are five so-called Platonic solids; the cube, and the regular figures called the tetrahedron, the octahedron, the icosahedron and the dodecahedron. The enthusiastic reader should repeat the above analysis for the other Platonic solids. Templates for constructing paper models of the Platonic solids are available at the book's web site.

Example 1.17

The integers $(\mathbb{Z}, +)$ form a group under addition. However, we are trying to persuade the reader that groups arise in nature as collections of bijections. To this end we view $\mathbb{Z}$ as a metric space. We do this by thinking of $\mathbb{Z}$ as a set equipped with a distance function which happens to satisfy the metric space axioms. If you are not familiar with metric spaces, it does not really matter. The point is that there is a sensible notion of the distance between each pair of elements of $\mathbb{Z}$. The distance function d in this instance is defined by $d(x, y) = |x - y|$. For each $z \in \mathbb{Z}$ we have an isometric (distance preserving) bijection $\text{add}_x : \mathbb{Z} \to \mathbb{Z}$ defined by $z \mapsto x + z$. Geometrically, this map shifts points a distance of x to the right if x is non-negative or a distance of $-x$ to the left if x is negative. The distances between points are preserved by the application of such maps. Notice that if $x, y \in \mathbb{Z}$ and $x \neq y$, then $\text{add}_x \neq \text{add}_y$, and that for all $u, v \in \mathbb{Z}$ we have $\text{add}_u \circ \text{add}_v = \text{add}_{u+v}$ and so the group $\{\text{add}_x \mid x \in \mathbb{Z}\}$ under function composition is a copy of $(\mathbb{Z}, +)$. In this case it does not matter if we are composing functions from right to left or left to right, because the group under discussion is commutative.

1.3 Cosets

Suppose that G is a group, and $A, B \subseteq G$. We define the product of subsets in the obvious way by $AB = \{ab \mid a \in A, b \in B\}$. Note that if either A or B is the empty set, then so too is the product. Equally well, if A (respectively B) is non-empty, and $G = B$ (respectively $G = A$), then $AB = G$. To see this, note that $AB \subseteq G$ by definition. Now we show that $G \subseteq AB$. Since $A \neq \emptyset$, we may choose $a \in A$. If $g \in G$ is chosen arbitrarily, then $g = a(a^{-1}g) \in AB$ and we are done.

If H is a subgroup of G, then $HH = H$ by the closure property. However, the converse is false. For example, $\emptyset\emptyset = \emptyset$ but $\emptyset \not\leq G$, since $\emptyset$ contains no elements at all, and so cannot contain an identity element! More seriously, consider $P = \{r \mid r \in \mathbb{R}, r \geq 1\}$ under multiplication. We have $PP = P$ but the circular constant π has no inverse in P.

Notice that multiplication of subsets of G is an associative operation. This is because for any $A, B, C \subseteq G$ we have

$$
\begin{aligned}
(AB)C &= \{(ab)c \mid a \in A, b \in B, c \in C\} \\
&= \{a(bc) \mid a \in A, b \in B, c \in C\} = A(BC).
\end{aligned}
$$

Thus parentheses are irrelevant.

A particularly interesting situation when forming a product AB is when one of the two sets is a subgroup, and the other is a singleton set. Thus the two sets involved are $H \leq G$ and $\{x\}$ for some $x \in G$. It quickly becomes tiresome to write $H\{x\}$ and $\{x\}H$, so we write Hx and xH instead. We call such subsets *right cosets* and *left cosets* of H in G (respectively).

We will discuss right cosets, but an entirely similar tale can be told for left cosets.

Lemma 1.18

Suppose that H is a subgroup of G.

(i) Suppose that $x \in G$, then $Hx = H$ if and only if $x \in H$.

(ii) Suppose that $x, y \in G$, then $Hx = Hy$ if and only if $xy^{-1} \in H$.

(iii) Suppose that $x \in G$, then $Hx = \{g \mid Hx = Hg\}$.

(iv) Suppose that $x \in G$, then $x \in Hx$.

Proof

(i) If $Hx = H$, then $x = 1x \in Hx = H$ so $x \in H$. Conversely, if $x \in H$, then $Hx \subseteq H$ because H is closed under multiplication. It remains to show that $H \subseteq Hx$. Choose any $h \in H$, then $h = (hx^{-1})x \in Hx$ because $h, x^{-1} \in H$. Thus $H \subseteq Hx$ and so $H = Hx$.

(ii) Notice that if $Hx = Hy$, then $Hxy^{-1} = Hyy^{-1}$ so $Hxy^{-1} = H$. On the other hand, if $Hxy^{-1} = H$, then $Hxy^{-1}y = Hy$ so $Hx = Hy$. Now apply part (i).

(iii) If $g \in Hx$, then $g = hx$ for some $h \in H$, so $Hg = Hhx = Hx$. Therefore $Hx \subseteq \{g \mid Hx = Hg\}$. Conversely if $Hx = Hg$, then $g = 1 \cdot g \in Hg = Hx$ so $\{g \mid Hx = Hg\} \subseteq Hx$. We are done.

(iv) This is a triviality, since $x = 1 \cdot x \in Hx$.

□

Part (ii) of Lemma 1.18 gives a criterion for deciding when two (right) cosets of H in G are equal. If two right cosets of H in G are not equal, then something very interesting happens.

Lemma 1.19

Suppose that H is a subgroup of G, and $x, y \in G$. It follows that either $Hx = Hy$ or $Hx \cap Hy = \emptyset$.

Proof

Either $Hx \cap Hy = \emptyset$ or there exists $z \in Hx \cap Hy$ in which case $zx^{-1}, zy^{-1} \in H$. Thus $xy^{-1} = (zx^{-1})^{-1}zy^{-1} \in H$ so by Lemma 1.18(ii) we have $Hx = Hy$.

□

Thus a pair of right cosets of H in G are either disjoint or equal.

Definition 1.20

Suppose that H is a subgroup of G. Let

$$G/H = \{xH \mid x \in G\} \text{ and } H\backslash G = \{Hx \mid x \in G\}.$$

Notice that if $g \in G$, then $g = g \cdot 1 = 1 \cdot g$ so $g \in gH$ and $g \in Hg$. Thus

$$G \subseteq \bigcup_{g \in G} Hg \leq G \text{ and } G \subseteq \bigcup_{g \in G} gH \leq G.$$

Thus each of G/H and $H\backslash G$ is a partition of G (a collection of non-empty subsets of G which are pairwise disjoint and have union which is G).

A *transversal* for a partition is a set which contains exactly one element from each set comprising the partition. In loose language, a transversal is a selection of a representative from each set in the partition.

Definition 1.21

We continue to assume that $H \leq G$. A transversal T for the partition $H \backslash G$ of G is called a set of *right coset representatives* for H in G (alternatively T can be called a *right transversal* for H in G).

Every right coset will be expressible as Ht for a unique $t \in T$. Note that the set T is in a natural bijective correspondence with $H \backslash G$ via $t \mapsto Ht$. There is an entirely similar notion of a set S of *left coset representatives* also known as a *left transversal* for H in G.

Suppose that G is a group. Inversion $x \mapsto x^{-1}$ is a bijection from G to G. Inversion induces maps $G/H \to H \backslash G$ and $H \backslash G \to G/H$ each using the recipe $X \mapsto X^{-1}$ where $X^{-1} = \{x^{-1} \mid x \in X\}$. We must check a few details. If $g \in G$, then

$$(gH)^{-1} = \{(gh)^{-1} \mid h \in H\} = \{h^{-1}g^{-1} \mid h \in H\} = \{hg^{-1} \mid h \in H\} = Hg^{-1},$$

and similarly $(Hg)^{-1} = g^{-1}H$. Thus the definition of our maps make sense in view of the fact that the set made by inverting all elements of a left (right) coset is a right (left) coset. The maps are mutually inverse bijections between G/H and $H \backslash G$. Note that if T is a left transversal for H in G (i.e. a set of left coset representatives), then it follows from our analysis that T^{-1} is a right transversal for H in G. Thus G/H and $H \backslash G$ have the same cardinality (size). This gives us the opportunity to make an important definition.

Definition 1.22

Let H be a subgroup of G. The cardinality of G/H (or equivalently $H \backslash G$) is called the *index* of H in G, and is written $|G : H|$.

Notice that $|G : H| = 1$ if and only if $H = G$, and that $|G : 1| = |G|$. These simple observations are as much at the heart of group theory as was Lemma 1.1. Please reflect on them.

We give a symbolic picture of a coset decomposition in Figure 1.1, where there are deemed to be 8 left cosets of H in G. The set of elements of G is to be thought of as the interior of the large rectangle, and each left coset H is to be thought of as the interior of a thin vertical rectangle. The particular elements in the left transversal are indicated by dots. Note that the number of vertical rectangles is the index of H in G, and in general this may be any positive integer or it may be infinite.

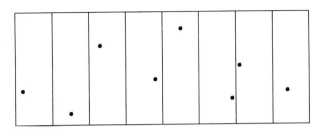

Figure 1.1 G is partitioned into 8 left cosets, and a transversal selected

Remark 1.23

In the course of the paragraph following Definition 1.21 we used the adjective *natural* to indicate an attribute of a particular bijective map. We smuggled in the term without a formal definition because we are using the term *natural* in a natural way. Let us be explicit; we say that a map is natural if it is the uniquely obvious or distinguished one. There is a topic called *category theory* in which the word *natural* has a precisely defined mathematical meaning. This is not a book about category theory, and we use the term in an informal way. This does not cause too many problems, since the category theoretic notion of *natural* is a refinement of our casual use.

Note that cosets were defined in a purely algebraic manner. In order to build confidence we will give some geometric examples. The geometry is not logically necessary, but many people find visual examples attractive. The first three examples are all in an abelian context, but this is no real drawback because we illustrate left and right cosets at the same time!

Example 1.24

Let $\mathbb{R}^* = \mathbb{R} - \{0\}$ be the group of non-zero real numbers under multiplication. Geometrically we think of the real line with 0 excised. Consider the subgroup $\mathbb{R}^+$ of positive real numbers. Now $\mathbb{R}^- = \mathbb{R}^* - \mathbb{R}^+$ is the set of negative real numbers. If $r > 0$ we have $\mathbb{R}^+ r = \mathbb{R}^+$ and if $r < 0$ we have $\mathbb{R}^+ r = \mathbb{R}^-$. Note that $\mathbb{R}^+ \cup \mathbb{R}^- = \mathbb{R}^*$ and $\mathbb{R}^+ \cap \mathbb{R}^- = \emptyset$. Thus there are two right cosets of $\mathbb{R}^+$ in $\mathbb{R}^*$.

Since $\mathbb{R}^*$ is an abelian group, there is no distinction between left and right cosets. A set of (left or right) coset representatives for $\mathbb{R}^+$ in $\mathbb{R}^*$ is $\{1, -1\}$. Another one is $\{-\sqrt{2}, \pi\}$. Alternatively we can say that both $\{1, -1\}$ and $\{-\sqrt{2}, \pi\}$ are left (or right) transversals for $\mathbb{R}^+$ in $\mathbb{R}^*$.

Example 1.25

Let $\mathbb{C}^* = \mathbb{C} - \{0\}$ be the (abelian) group of non-zero complex numbers under multiplication. This has a geometric interpretation consisting of the complex plane punctured at the origin. Let

$$S = \{z \mid z \in \mathbb{C}^*, |z| = 1\} - \{e^{i\theta} \mid \theta \in \mathbb{R}\}.$$

Note that S is a subgroup of $\mathbb{C}^*$. We investigate the cosets of S in $\mathbb{C}^*$. Suppose that $w \in \mathbb{C}^*$ so $w = re^{i\psi}$ for some real ψ and some positive real r. Now

$$Sw = \{e^{i\theta}re^{i\psi} \mid \theta \in \mathbb{R}\} = \{re^{i\varphi} \mid \varphi \in \mathbb{R}\}.$$

Thus the cosets of S, viewed geometrically in the Argand diagram, are the circles centred at the origin. Notice that any pair of distinct such circles have empty intersection, and the union of all such circles is the complex plane punctured at the origin as we should expect. There are infinitely many cosets of S in $\mathbb{C}^*$. An obvious set of (left or right) coset representatives for S in $\mathbb{C}^*$ is is $\mathbb{R}^+$ as discussed in Example 1.24. Another (left or right) transversal for S in $\mathbb{C}^*$ is $\mathbb{R}^-$ also drawn from Example 1.24. These transversals are respectable and are likely to be easy to deal with mathematically. However, wild choices of coset representatives are available. *Skip to the next example if you don't know about cardinality.* Let c be the cardinality of the reals. If you restrict options for selecting coset representatives by insisting that each representative is real, you still have to choose between r and $-r$ as a coset representative for every positive real number r. There are c positive real numbers and so 2^c such transversals (i.e. the cardinality of the power set of the reals). This cardinality is strictly larger than c.

Example 1.26

Consider the group $\mathbb{R}^2$ under addition. In the usual geometrical interpretation, the elements of $\mathbb{R}^2$ are thought of as points in 2-dimensional Euclidean space, and addition of points corresponds to vector addition of their position vectors. Any straight L line through the origin will be a subgroup, and its cosets will be the family of straight lines which are parallel to it. Given a pair of parallel straight lines, either they are the same straight line, or they are disjoint. As in the previous example, there will be a prodigous infinity of bizarre transversals. However, mathematically pleasant ones are readily available. For example, take any line L' which is not parallel to L. The points of L' will form a (left or right) transversal for L in $\mathbb{R}^2$.

A similar example is obtained by considering a plane through the origin in $\mathbb{R}^3$. Any such plane P will form an additive subgroup, and its cosets will be the planes which are parallel to it. Again, two parallel planes in $\mathbb{R}^3$ either coincide or are disjoint. As usual we seek beautiful transversals; any straight line which intersects P in a singleton set will do nicely.

Example 1.27

Look back to Example 1.16. The full group C of isometries of a cube has size 48, but only 24 of the isometries can be accomplished by using a rigid motion in 3-dimensional space. These 24 rigid motions comprise a subgroup D. There are two cosets of D in C : one is D itself, and the other is $C - D$. A mirror may help! This is true whether you are working with left cosets or right cosets. Suppose that $\{x, y\}$ is a left transversal for D in C. We may suppose that $x \in D$ and $y \in C - D$. Now $xD = D = Dx$ and $yD = C - D = Dy$. Thus even though C is not an abelian group, $\{x, y\}$ also turns out to be a right transversal for D in C. Since x, y were arbitrary choices from their respective left cosets of D in C the equations $xD = Dx$ and $yD = Dy$ also force D to have the property that $gD = Dg$ for every $g \in C$. We will develop this important idea in Section 2.2.

Finally we look at a specific subgroup of a non-abelian group in order to see how left cosets and right cosets may differ. We have to leave aside those engaging geometric examples to find a group which is sufficiently tame that it is easy to see what is happening, but the group structure is wild enough so that left and right cosets differ.

Example 1.28

Recall Example 1.13 where we showed that there was a group

$$S = \{e, \alpha, \beta, \gamma, \delta, \varphi\}$$

with the following multiplication table:

	e	α	β	γ	δ	φ
e	e	α	β	γ	δ	φ
α	α	e	φ	δ	γ	β
β	β	δ	e	φ	α	γ
γ	γ	φ	δ	e	β	α
δ	δ	β	γ	α	φ	e
φ	φ	γ	α	β	e	δ

(1.6)

Let $H = \{e, \alpha\}$. The left cosets of H in S are

$$\{e, \alpha\}, \{\gamma, \varphi\}, \{\beta, \delta\}.$$

The right cosets of H in S are

$$\{e, \alpha\}, \{\gamma, \delta\}, \{\beta, \varphi\}.$$

Thus e, γ, δ is a left transversal but not a right transversal, whereas e, γ, φ is a right transversal but not a left transversal. However e, γ, β is simultaneously a left transversal and a right transversal for H in S.

One of the phenomena which arises when examing Example 1.28 is an instance of a result of Philip Hall which we will not prove here. It asserts that when G is a group and H is a subgroup of G of finite index, it is always possible to choose a set of left coset representatives (a left transversal) which is simultaneously a set of right coset representatives (a right transversal). The interested reader can look for the result known as *Hall's marriage theorem* which gives necessary and sufficient conditions for a set of n women to be able to marry (monogamously and simultaneously) n men in such a way that at least the women are initially happy. In what passes for group theoretic reality, the women are the left cosets of H in G and the men are the right cosets of H in G. The details of the proof and its application are available at this book's web site.

Lemma 1.29

Suppose that G is a group with subgroups K and L such that $K \leq L \leq G$. It follows that $|G : L| \times |L : K| = |G : K|$.

Before we begin the proof, we observe that it follows from Lemma 1.29 that if at least one of $|G : L|$ and $|L : K|$ is infinite, then so too is $|G : K|$. Moreover, the converse statement is also true. Our proof will even show that the equation is valid when it is an equality of infinite cardinals, but the reader who does not know about such matters should not panic. There are notes about cardinality available on the internet at

http://www.bath.ac.uk/~masgcs/book1/amplifications/ch1.html

suitable for a multi-faith readership. The topic is also covered in a wide range of books.

Proof

Let A be a left transversal for L in G and let B be a left transversal for K in L. Suppose that $a, a' \in A$, $b, b' \in B$ and that $abK = a'b'K$. It follows that $abKL = a'b'KL$ so $abL = a'bL$ and therefore $aL = a'L$ since $b, b' \in L$. Now

$a, a' \in A$, a left transversal for L in G so $a = a'$. Therefore $bK = b'K$. Now B is a left transversal for K in L, so $b = b'$. In particular, it follows that $ab = a'b'$ if and only if $a = a'$ and $b = b'$.

We now know that $|AB| = |A| \times |B|$ (and this may be an equality of infinite cardinals). It remains to show that

$$G = \bigcup_{c \in AB} cK.$$

However, $ABK = A(BK) = AL = G$ and we are done.

$\square$

Corollary 1.30

If there are $n \in \mathbb{N}$ subgroups $H_1, \ldots, H_n$ of G such that $H_1 \leq H_2 \leq \ldots \leq H_n$, then

$$|H_n : H_1| = \prod_{i=1}^{n-1} |H_{i+1} : H_i|.$$

Example 1.31

Let $G = \mathbb{R}^3$ under addition. Consider a subgroup P consisting of a plane through the origin, and a subgroup of P consisting of a straight line L through the origin. The cosets of L in $\mathbb{R}^3$ are all straight lines parallel to L. One can obtain a transversal for L in $\mathbb{R}^3$ (a set of coset representatives) by choosing a transversal T for L in P, and a transversal S for P in $\mathbb{R}^3$. Then as s and t range over S and T respectively, the elements $s + t$ will be distinct, and will comprise a transversal for L in $\mathbb{R}^3$. We do not have to distinguish left and right transversals since $\mathbb{R}^3$ is an abelian group.

Proposition 1.32

If a group G has subgroups H_i where $1 \leq i \leq n$, and $|G : H_i| < \infty$ for each i, then $\cap_{i=1}^n H_i$ has finite index in G.

Proof

Suppose that H, K are both subgroups of finite index in a group G. Let $T = H \cap K$. Choose a right transversal R for T in K (i.e. R is a set of right coset representatives for T in K). Suppose that $a, b \in R$. If $Ha = Hb$ then $ab^{-1} \in H \cap K$ so $(H \cap K)a = (H \cap K)b$ and therefore $a = b$. It follows that the map $(H \cap K)\backslash K \to H\backslash G$ defined by $(H \cap K)x \mapsto Hx$ $(x \in R)$ is injective,

However, $H \backslash G$ is a finite set (of size $|G : H|$) and so R is finite, and therefore $|K : T| < \infty$. Now

$$|G : H \cap K| = |G : T| = |G : K||K : T| < \infty$$

and we are done.

Next we address the general case where there are n subgroups H_i each of finite index in G. We prove that $S = \cap_{i=1}^{n} H_i$ has finite index in G by induction on n, the case $n = 1$ being trivial, and the case $n = 2$ having just been done. Thus we may assume that $n > 2$. Put $R = \cap_{i=1}^{n-1} H_i$ so $|G : R| < \infty$ by inductive hypothesis. Now $|G : R \cap H_n| < \infty$ by the case $n = 2$, but $R \cap H_n = S$ and so we are done.

$\square$

Next we obtain perhaps the most well-known result in the theory of groups.

Theorem 1.33 (Lagrange)

Suppose that G is a finite group, and that H is a subgroup of G, then

$$|G| = |G : H| \times |H|.$$

Proof

Since $|H| = |H : 1|$, this is a special case of Lemma 1.29.

$\square$

Another way to see this is to observe that there are $|G : H|$ left cosets of H in G, and that each left coset has exactly $|H|$ elements because the map $h \mapsto xh$ is a bijection between H and xH (with inverse $y \mapsto x^{-1}y$ for every $y \in xH$). Now G is the union of $|G : H|$ disjoint sets, each of size $|H|$, so the theorem follows.

Suppose that $H, K \leq G$. In general the subset HK is not a subgroup of G. However, it is a rather interesting set, not least because it is simultaneously a union of right cosets of H and a union of left cosets of K. To be explicit, we have

$$HK = \bigcup_{k \in K} Hk = \bigcup_{h \in H} hK.$$

We note that for $k_1, k_2 \in K$ we have $Hk_1 = Hk_2$ if and only if $k_1 k_2^{-1} \in H$. This is the same condition for the cosets $(H \cap K)k_1, (H \cap K)k_2 \in (H \cap K) \backslash K$ to coincide. Thus the number of sets of the form Hk as k ranges over K is

$|K : H \cap K|$. The sets Hk are each in bijective correspondence with H, so if H, K are finite we have

$$|HK| = |H||K : H \cap K| = \frac{|H| \cdot |K|}{|H \cap K|}. \tag{1.7}$$

This is an important formula, and we will return to it later.

1.4 Subgroup Generation

Suppose that G is a group and that $X \subseteq G$. The subgroup generated by X is the smallest subgroup of G which contains X. That was a touch informal, so we supply a formal definition to be on the safe side.

Definition 1.34

Suppose that G is a group and that $X \subseteq G$. Let $S_X = \{H \mid H \leq G, X \subseteq H\}$. Observe that $S_X \neq \emptyset$ since $G \in S_X$. The subgroup $\langle X \rangle$ generated by X is defined to be

$$\langle X \rangle = \bigcap_{H \in S_X} H. \tag{1.8}$$

Recall that Proposition 1.12 asserts that the intersection of subgroups is a subgroup.

This group $\langle X \rangle$ is therefore the unique smallest subgroup of G which contains X. Note that for $H \subseteq G$ we have $H \leq G$ if and only if $H = \langle H \rangle$.

A *word* on X is any finite sequence $w(X)$ of elements of X and their inverses. Suppose that $x_1, x_2, \ldots, x_n$ is such a sequence of length n. We define the value of the word to be s_n where $s_0 = 1$ and $s_r = s_{r-1} x_r$ when $r > 0$. Casually you might say that the value of a word is obtained by multiplying the terms together in the specified order. However, by being a little more careful we have given meaning to the values of words of length 0 and 1 without having to single them out for special treatment. Thus the empty word (of length 0) has value 1 and a product involving just one x or x^{-1} is itself. When writing a word (a finite sequence) in this context it is usual to omit the commas. Thus if $x, y, z \in X$ then $xy^{-1}yz^{-1}x$ is a word, as are $1, x$ and x^{-1}.

The set of values of words on X is manifestly closed under multiplication. It is also closed under inversion (reverse the word and change the sign of the exponents) and contains X. Thus it forms a subgroup of G. Moreover, if $H \leq G$ and $X \subseteq H$, then our group of word values is contained in H. Thus this collection of word values is the same as $\langle X \rangle$. We now have two characterizations of $\langle X \rangle$, one external, and one internal. This is very useful, since in any given situation, one perspective is likely to be clearer than the other.

So far we have been careful to distinguish between a word and its value. In future we will assume that the distinction is understood, and will allow ourselves to be sloppy when confusion seems unlikely. When studying a group, it is usually very instructive to understand its subgroups. The intersection of subgroups is itself a subgroup, but in general the union of subgroups is not a subgroup. Suppose that A and B are subgroups of G, we write $\langle A, B \rangle$ for the intersection of all the subgroups of G which contain both A and B. Note that G itself contains both A and B, so we are not intersecting a dangerously empty family of groups! Thus $\langle A, B \rangle$ is the smallest subgroup of G which contains both A and B and in fact $\langle A, B \rangle$ is just another name for $\langle A \cup B \rangle$. We refer to $\langle A, B \rangle$ as the *join* of A and B.

A *Hasse diagram* is a picture of at least part of the family of subgroups of a group. A vertical line, or a line sloping up from a dot labelled by a subgroup to another dot labelled by a subgroup indicates inclusion. A label on a line indicates the index. Two lines coming down from different subgroups should meet with the label of the intersection of the two subgroups, and lines coming up from distinct subgroups should meet at the join of the two subgroups.

In Figure 1.2 we give diagrams showing all the subgroups of two particular groups B and V. Now B consists of the four given complex numbers under multiplication, and it is easy to see that B has exactly three subgroups $1, A$ and B as shown on the left of Figure 1.2. Notice that $2 \times 2 = 4 = |B|$.

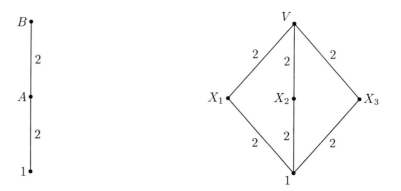

$$A = \{1, -1\}, \; B = \{\pm 1, \pm i\} \qquad\qquad \text{A group } V \text{ of order } 4$$

Figure 1.2 Hasse diagrams of two groups of order 4

The second group illustrated in Figure 1.2 consists of four 2×2 matrices with entries which are integers. Here I is the 2×2 identity matrix and α differs from I only in that the top left entry is -1. Thus $V = \{I, -I, \alpha, -\alpha\}$ is a group under matrix multiplication. Note that the square of any element of V is I and the product of any pair of distinct non-identity elements is the third non-identity element. This group (and groups which are just copies of it) is quite well-known, and is called the Vierergruppe (literally "foursgroup"). This is why we have called this group V. It has exactly 5 subgroups; itself, the trivial subgroup, and three subgroups of order 2. The intersection of any pair of distinct subgroups of order 2 is the trivial group, and the join of any pair of distinct subgroups of order 2 is V. All this information is summarized in the diagram on the right of Figure 1.2.

Some group theorists find it very convenient to think about groups in terms of these diagrams. In particular, the isomorphism theorems which we will prove in Chapter 2 have very pleasant interpretations in terms of Hasse diagrams.

EXERCISES

1.13 Suppose that G is a group and that A, B are subgroups of G. Show that AB is a subgroup of G if and only if $AB = BA$.

1.14 Suppose that G is a group containing subgroups H and K. Show that $H \cup K = \langle H, K \rangle$ if and only if $H \cap K \in \{H, K\}$.

1.15 Show that there is a group G containing subgroups H, K, L such that $H \cup K \cup L = \langle H, K, L \rangle$ but $H \cap K \cap L \notin \{H, K, L\}$.

1.5 Finite Generation

A group G is said to be *finitely generated* if there is a finite set X such that $G = \langle X \rangle$. Every finitely generated group is countable, since there are only finitely many elements of G which are expressible using a word of length n in the generators and their inverses, and the countable union of finite sets is countable. On the other hand, not every countable group is finitely generated. The rational numbers $\mathbb{Q}$ under addition are a case in point, as we will see in Proposition 1.44.

Observe that if G is countably generated (so $G = \langle X \rangle$ where X is a countable set), then G is countable by a slightly polished version of the argument in the previous paragraph. This time we need to know that the union of countably many countable sets is countable.

EXERCISES

1.16 Suppose that H is a subgroup of G, and that H is finitely generated. We are also given that the index $|G : H|$ is finite. Show that G is finitely generated.

1.17 Suppose that G is a finitely generated group and that $G = \langle Y \rangle$ where Y is not necessarily a finite set. Prove that there is a finite subset X of Y such that $G = \langle X \rangle$.

The trivial group needs no generators (or if you prefer, the empty set generates the trivial group). Those groups which can be generated by a single element will obviously attract our attention.

Definition 1.35

A group G is *cyclic* if there is $x \in G$ such that $G = \langle \{x\} \rangle$.

Everyone actually writes $\langle x \rangle$ rather than $\langle \{x\} \rangle$, and we allow ourselves to say that x generates $\langle x \rangle$ (rather than the accurate but ridiculously pedantic "$\{x\}$ generates $\langle \{x\} \rangle$"). In fact if $X = \{x_1, \ldots, x_n\}$ is a finite subset of a group, it is quite acceptable to write $\langle x_1, x_2, \ldots, x_n \rangle$ instead of $\langle X \rangle$. You can say that X generates $\langle X \rangle$, or that $x_1, x_2, \ldots, x_n$ are generators for $\langle X \rangle$.

Example 1.36

Here are some cyclic groups.

(i) The subset $\{2^i \mid i \in \mathbb{Z}\}$ of $\mathbb{R}$ under multiplication. These elements form the set

$$\{\ldots, 1/8, 1/4, 1/2, 1, 2, 4, 8, \ldots\}$$

and either 2 or 1/2 can be used as a generator.

(ii) The subset $\{-1, 1\}$ of $\mathbb{R}$ under multiplication. In this case only -1 is a generator.

(iii) The subset $\{e^{2\pi ki/12} \mid k \in \mathbb{Z}\}$ of $\mathbb{C}$ under multiplication. This set consists of the 12 complex numbers $e^{2\pi ki/12}$ where k ranges from 0 to 11 (other values of k merely repeat numbers we have listed). See Figure 1.3. Our complex numbers all have modulus 1, and so in the Argand diagram they are located on the unit circle centred at the origin. These points in the Argand diagram comprise the vertices of a regular 12-sided polygon. Notice that $1, i, -1$ and i are among the complex numbers in question. Let $\lambda = e^{2\pi i/12}$, so $\lambda^3 = i$ and $\lambda^6 = -1$. Please verify that $\langle \alpha \rangle = \langle \lambda \rangle$ if and only if $\alpha \in \{\lambda, \lambda^5, \lambda^7, \lambda^{11}\}$. Incidentally, note that $\lambda^{-1} = \bar{\lambda} = \lambda^{11}$.

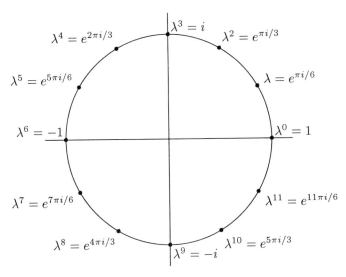

Figure 1.3 The twelve complex numbers λ which are roots of $X^{12} - 1$

(iv) The set of 2×2 matrices

$$\left\{ \begin{pmatrix} 1 & x \\ 0 & 1 \end{pmatrix} \middle| x \in \mathbb{Z} \right\}$$

under multiplication. There are two possible generators for this group. Please find them.

Proposition 1.37

Let G be a group of prime order. It follows that G is cyclic.

Proof

Choose $x \in G$ with $x \neq 1$. Let $H = \langle x \rangle$ be the cyclic group generated by x. Now $H \neq 1$ so by Lagrange's theorem it must be that $|H| = p$.

□

Proposition 1.38

Every cyclic group $\langle x \rangle$ is abelian.

Proof

This follows from the definition of a cyclic group and Corollary 1.9.

$\square$

Definition 1.39

Suppose that G is a group and that $x \in G$. We define $o(x)$, the order of x, to be $|\langle x \rangle|$.

Proposition 1.40

Suppose that $G = \langle x \rangle$ is a cyclic group. It follows that exactly one of the following two statements is true.

(i) $|G| = n < \infty$, in which case the elements $1 = x^0, x^1, x^2, \ldots, x^{n-1}$ are distinct, $G = \{x^0, x^1, \ldots, x^{n-1}\}$ and $x^n = 1$.

(ii) $|G|$ is infinite, in which case the elements x^i, $i \in \mathbb{Z}$ are distinct.

Proof

Suppose that there are $i, j \in \mathbb{Z}$ with $i < j$ and $x^i = x^j$ (if there are no such i and j we are in case (ii)). Among all such i and j choose a pair with $j - i = n$ as small as possible. Now $x^i = x^j$ so $1 = x^{-i}x^i = x^{-i}x^j = x^n$. Moreover, $x^0, x^1, \ldots, x^{n-1}$ must be distinct otherwise we would violate the choice of n. Finally, if $a \in \mathbb{Z}$ we can find $q, r \in \mathbb{Z}$ with $0 \le r < n$ such that $a = qn + r$. Now $x^a = x^{qn+r} = (x^n)^q x^r = 1^q x^r = x^r$. Thus $G = \{x^0, x^1, \ldots, x^{n-1}\}$ and case (i) is established. Our proof is complete.

$\square$

Corollary 1.41

Let G be a group and suppose $x \in G$ has finite order n. Suppose that m is an integer. We have $x^m = 1$ if and only if n divides m.

Remark 1.42

Thus if $x \in G$ has finite order n, then n is the least positive integer such that $x^n = 1$ and Proposition 1.40 applies.

Proposition 1.43

If $G = \langle x \rangle$ is a cyclic group, and $H \leq G$, then H is a cyclic group.

Proof

If $H = 1$ there is nothing to prove, so we may assume that $H \neq 1$. For some $n \in \mathbb{Z} - \{0\}$ we have $x^n \in H$. Among all such n, choose one which minimizes $|n|$. Certainly $\langle x^n \rangle \leq H$. Next choose an arbitrary $h \in H$. Now $h = x^a$ for some $a \in \mathbb{Z}$ since G is cyclic with generator x. Now $a = qn + r$ for $q, r \in \mathbb{Z}$, with $0 \leq r < |n|$. Now $h(x^n)^{-q} = x^r \in H$. The choice of n forces $r = 0$, so $h \in \langle x^n \rangle$ and we have $H \leq \langle x^n \rangle$. However, we already know that $\langle x^n \rangle \leq H$ so $H = \langle x^n \rangle$ is cyclic.

$\square$

Proposition 1.44

The additive group $\mathbb{Q}$ of rational numbers is not finitely generated.

Proof

Suppose (for contradiction) that $\mathbb{Q}$ is generated by finitely many rational numbers q_i where $i = 1, 2, \ldots m$. Choose a positive integer n such that $nq_i \in \mathbb{Z}$ for every i. Let $A = \langle 1/n \rangle$. Now $q_i \in A$ for every i, so $\mathbb{Q} \leq A \leq \mathbb{Q}$. We conclude that $\mathbb{Q} = A$ but $1/2n \notin A$ so this is absurd. The proof is complete.

$\square$

Proposition 1.45

Suppose that $G = \langle x \rangle$ is a cyclic group.

(i) If G is infinite, then G has exactly one subgroup of index n for each $n \in \mathbb{N}$, and exactly one subgroup of infinite index.

(ii) If G is finite of order k, then G has exactly one subgroup of each order which is a natural number divisor of k, and no others.

Proof

(i) Let $n \in \mathbb{N}$, then $\langle x^n \rangle = \langle x^{-n} \rangle$ has index n, since the n cosets $\langle x^n \rangle x^i$ for $0 \leq i < n$ are distinct, and their union is $|G|$. Since a subgroup of a cyclic group is cyclic by Proposition 1.43, the only other subgroup of G is $\langle x^0 \rangle = 1$ which has infinite index in $|G|$.

(ii) If u is a natural number such that u does not divide k, then there is no subgroup of order u by Lagrange's Theorem. Instead suppose that u divides k (we may write this as $u \mid k$). The subgroup $\langle x^{k/u} \rangle$ has order u. Next we tackle the uniqueness question. Suppose that $H \leq G$ and $|H| = u$. If $u = 1$ there is no uniqueness problem, so we may assume that $u > 1$. Now $H = \langle x^t \rangle$ for some $t \in \mathbb{N}$. By Lagrange's Theorem we have $x^{tu} = 1$ so $k \mid tu$ and thus $k/u \mid t$. We conclude that x^t is a power of $x^{k/u}$ so $H \leq \langle x^{k/u} \rangle$. However, the group $\langle x^{k/u} \rangle$ has order u, as does H, so $H = \langle x^{k/u} \rangle$ and we are done.

$\square$

If $\langle x \rangle$ is an infinite cyclic group, and j is a non-zero integer, then x^j has infinite order. However, the situation for finite cyclic groups is more subtle.

Proposition 1.46

Suppose that $G = \langle x \rangle$ is a cyclic group of order $n < \infty$. Suppose that $j \in \mathbb{Z}$, then the order of x^j is $n/\text{g.c.d.}(n, j)$.

Proof

First note that $\mu = n/\text{g.c.d.}(n, j)$ and $\nu = j/\text{g.c.d.}(n, j)$ are coprime. Now for $\lambda \in \mathbb{N}$ we have $(x^j)^\lambda = 1$ if and only if n divides λj by Corollary 1.41. We rephrase this condition: $n = \mu \text{g.c.d.}(n, j)$ divides $\lambda j = \lambda \nu \text{g.c.d.}(n, j)$. This is equivalent to $\mu \mid \lambda$, which just asserts that λ is an integer multiple of $n/\text{g.c.d.}(n, j)$. Thus the order of x^j is $n/\text{g.c.d.}(n, j)$.

$\square$

Notice that $\mathbb{Z}$ is an additive infinite cyclic group. There are several rival notations for a generic additive cyclic group of order n, one of which is $\mathbb{Z}_n$, another is $\mathbb{Z}_{[n]}$, and yet another is $\mathbb{Z}/n\mathbb{Z}$. This is all a bit much really, but happily there is settled notation when writing generic cyclic groups multiplicatively. The infinite cyclic group is written C_∞, and the cyclic group of order n is written C_n.

Let C denote the set of complex numbers z with the property that for some natural number n_z we have $z^{n_z} = 1$ (here n_z is allowed to vary as z varies). We first show that C is a group under multiplication of complex numbers. For each natural number n let $C[n] = \{z \mid z \in \mathbb{C}, z^n = 1\}$. Note that $C[n]$ is a cyclic group with $e^{2\pi i/n}$ as a generator. This follows from the theory of complex numbers. Moreover, if n_1 divides n_2 then $C[n_1] \le C[n_2]$. Note that $C = \cup_{n \in \mathbb{N}} C[n]$. It now follows easily that C is a group because $1 \in C \ne \emptyset$, and if $x, y \in C$, then $x \in C[n_1], y \in C[n_2]$ for some $n_1, n_2 \in \mathbb{N}$. Now $x, y \in C[n_1 n_2]$, so $xy^{-1} \in C[n_1 n_2] \le C$.

In terms of $re^{i\theta}$ notation, we have

$$C[n] = \{e^{2\pi i\theta} \mid \theta \in \mathbb{Q}, n\theta \in \mathbb{Z}\}.$$

and

$$C = \{e^{2\pi i\theta} \mid \theta \in \mathbb{Q}, n\theta \in \mathbb{Z} \text{ for some } n \in \mathbb{N}\} = \{e^{2\pi i\theta} \mid \theta \in \mathbb{Q}\}.$$

There are groups G with the property that every proper subgroup of G is cyclic, but G is not cyclic. We now illustrate this point. Fix a prime number p. Let C_{p^∞} denote the set of complex numbers z with the property that for some natural number n_z we have $z^{n_z} = 1$ where $n_z = p^{i_z}$ for some integer $i_z \ge 0$ (and i_z is allowed to vary as z varies). We will show that C_{p^∞} is a group under multiplication of complex numbers. Furthermore we will demonstrate that C_{p^∞} is not a cyclic group, but that if H is a subgroup of C_{p^∞} with $H \ne C_{p^\infty}$ (i.e. H is a proper subgroup of C_{p^∞}), then H is a cyclic group. Notice that $C[p] \le C[p^2] \le \ldots$ and that $C_{p^\infty} = \cup_{i \in \mathbb{N}} C[p^i]$. The union of an ascending chain of subgroups is a group for routine reasons, so $C_{p^\infty} \le \mathbb{C}^*$. Moreover each $C[p^v]$ is a cyclic group of order p^v with generator $e^{2\pi i/p^v}$. In fact for $v \ge 1$ the elements of $C[p^v] - C[p^{v-1}]$ have order dividing p^v but not dividing p^{v-1}, and so have order p^v. It follows that each element of $C[p^v] - C[p^{v-1}]$ generates the cyclic group $C[p^v]$. If C_{p^∞} were cyclic with generator t, then $t \in C[p^m]$ for some $m \in \mathbb{N}$, so $C_{p^\infty} = \langle t \rangle \le C[p^m] \le C_{p^\infty}$. It would follow that $C_{p^\infty} = C[p^m]$ which is absurd since $C_{p^\infty} \ni e^{2\pi i/p^{m+1}} \notin C[p^m]$. Thus C_{p^∞} is not cyclic.

Now suppose that H is a proper subgroup of C_{p^∞}. Since $C_{p^\infty} = \cup_{i \in \mathbb{N}} C[p^i]$ there must exist a least $j \in \mathbb{N}$ with $C[p^j] \not\subseteq H$. Thus $C[p^{j-1}] \le H$. Since the elements of $C[p^v] - C[p^{v-1}]$ each generate the cyclic group $C[p^v]$, none of them can be in H. Thus $H \cap (C_{p^\infty} - C[p^{j-1}]) = \emptyset$ and so $H \cap C[p^j] = C[p^{j-1}]$. Now H contains no elements of order p^j, so $H \le C[p^{j-1}]$. We already knew that $C[p^{j-1}] \le H$ so $H = C[p^{j-1}]$ is cyclic (and finite).

Recall that we investigated cyclic groups because they arose as the most straightforward type of finitely generated group. If a group can be generated by a set of size n, then we say that it is an n-generator group. Thus 1-generator groups are just the cyclic ones. It might seem natural to move on to 2-generator groups, but this is a very wide class of groups indeed. Here are a few facts which should indicate to you how complicated 2-generator groups can be.

Remark 1.47

(a) Given any countable group G there is a 2-generator group H which contains a copy of G as a subgroup.

(b) There are groups called Tarski monsters, which were constructed by Ol'shanskii and Rips. Let p be a sufficiently large prime number ($p \geq 10^{40}$ will do nicely). There is a group T which has the property that it is infinite, and every subgroup is either T, the trivial group, or a copy of the cyclic group of order p. Thus any two elements $a, b \in T$ which do not lie in the same cyclic subgroup of order p have the property that $\langle a, b \rangle = T$.

(c) There is 2-generator group W which contains a subgroup which is not finitely generated.

We will try to understand some particularly straightforward 2-generator groups. Suppose that $G = \langle x, y \rangle$ and either $x \in \langle y \rangle$ or $y \in \langle x \rangle$. In this case we really have a 1-generator group and G must be cyclic. The easiest interesting case is when x and y are distinct and have order 2. This is an important class of groups so we make a definition.

Definition 1.48

A group D is said to be *dihedral* if it is generated by a pair of distinct elements x, y of order 2.

The name comes from the fact that the finite groups of this type arise as the groups of rigid symmetries of regular n-sided polygons, as will be clear to all readers who are fluent in Ancient Greek. This remark about rigid symmetries is commentary, and does not form part of our definition.

Let $z = xy$ so $z^{-1} = yx$. Notice that $z \neq 1$ since if $z = 1$ it follows that $x = x^{-1} = y$. Now z either has infinite order or order $n \in \mathbb{N}$ where $n > 1$. Suppose that $w \in G$, so w is a word on the letters x, y and their inverses. In fact $x^{-1} = x$ and $y^{-1} = y$ so w is a word on the letters x and y and no inverses are needed. Recall that $x^2 = y^2 = 1$ so we may erase any occurrence of a letter next to itself in w. Thus we may assume that w is either the identity or a word of one of the following types (depending on the first and last letters of w).

$$
\begin{array}{rcll}
xyxyx\ldots xy & = & z^m & m \geq 0 \\
xyxyx\ldots x & = & xz^{-m} & m \geq 0 \\
yxyxy\ldots yx & = & z^{-m} & m > 0 \\
yxyxy\ldots y & = & xz^m & m > 0 \quad \text{(insert } x^2 \text{ at the start)}
\end{array}
$$

Thus every element of G may be written as z^m or xz^m for some $m \in \mathbb{Z}$. Suppose, for contradiction, that $x \in \langle z \rangle$. We know $x \neq 1$ so $x = xyxyx\ldots xy$ or $x = yxy\ldots yx$. In either event we deduce an equation $yxyxy\ldots xy = 1$ which forces $y \in \langle z \rangle$. Now x and y are both elements of order 2 in the cyclic group $Z = \langle z \rangle$. By Proposition 1.45(ii) the group Z contains at most one subgroup of order 2, and that subgroup has exactly one non-identity element, so $x = y$ which is contrary to our assumptions. Thus $x \notin Z$, and similarly $y \notin Z$. We conclude that $|D : Z| = 2$. Thus if $n = o(z)$ is finite, then $|D| = 2n$. The elements of Z form a set of coset representatives for 1 in Z, so we may use Lemma 1.29 to describe the elements of D. Each element of D is either an element of Z, or xz where z is a uniquely determined element of Z.

In symbols we can list the elements of D (without repetition) as $x^i z^j$ where $i \in \{0,1\}$ and $j \in \mathbb{Z}$ if z has infinite order, or $0 \leq j < n$ if z has finite order n. Note that $zx = xz^{-1}$ and $z^{-1}x = xz$ so $z^j x = xz^{-j}$ for all $j \in \mathbb{Z}$. Thus we know how to multiply elements of D :

$$
x^{i_1} z^{j_1} x^{i_2} y^{j_2} = x^{i_1+i_2} z^{(-1)^{i_2} j_1 + j_2}
$$

where the arithmetic takes place mod 2 in the exponent of x, and mod n in the exponent of z if $n = o(z)$ is finite. If z has infinite order the arithmetic in the exponent of z takes place in the integers (or if you like, mod 0).

We have shown that if we have a group generated by distinct elements of order 2, then we can pin down its structure very precisely. We can describe its elements and its multiplication, and moreover we have candidate groups of each even order $2n \geq 4$, and one infinite candidate. Our language is a little cautious. After all, we do not know for certain that any of these groups exist! Perhaps (and this does not happen, but we have to worry) when we investigate these candidates further we might find something nasty lurking under a stone marked "the associative law". It might be that the candidate multiplication we have found is not associative if $n = 42$. We can easily see that our multiplication laws are closed, and identity elements and inverses exist, so associativity is the only possible problem.

Now, for each natural number $n \geq 3$, there are $2n$ rigid symmetries of a regular n-sided polygon. In such a figure, there are n axes of symmetry. Take a pair of symmetries which are reflections in adjacent axes of symmetry. Each has order 2, and their product is a rotation through $2\pi/n$. This pair of elements of order 2 therefore generate all $2n$ rigid symmetries of a regular n sided figure. Thus there are dihedral group of order $2n$ when n is finite and bigger than 2. The infinite case is very similar; the number line becomes the regular ∞-gon when we declare a vertex at each integer. We examined this figure in Example 1.17. Consider the following two rigid symmetries of this object: (a) reflection about 0 (the map sending the vertex at a to $-a$ for every $a \in \mathbb{Z}$, and (b) reflection about $1/2$ (the map sending the vertex at b to $-b + 1$. Their product is a translation by 1, so between them they generate the infinite group consisting of all the rigid symmetries of the regular ∞-gon. Thus the infinite dihedral group exists. There remains the case $n = 2$. Take a rectangle which is not square. Its group of rigid symmetries has size 4, and consists of the identity, reflections in each of two axes of symmetry which go through the mid-point and are parallel to a side, and a fourth element which is rotation through π about the mid-point. Any pair of distinct non-identity elements of this group qualify it for dihedral status, so there is a (slightly awkward) dihedral group of order 4. We know that these dihedral groups are essentially unique, because we can determine their multiplications. They therefore deserve names. We begin uncontroversially by calling the infinite one D_∞. The dihedral group of order $2n$ is sometimes called D_n and sometimes D_{2n}. We favour the latter for no good reason, save that $2n$ is the order of the group. Notice that D_6 is the same as S_3.

Elements of order 2 play a special role in many parts of group theory, and so have attracted special terminology. They are called *involutions*. Thus a dihedral group is a group generated by a pair of distinct involutions.

EXERCISES

Let D be a dihedral group generated by distinct involutions x and y. Let $z = xy$, and put $Z = \langle z \rangle$.

1.18 Suppose that $w \in D$ but $w \notin Z$. Prove that w is an involution.

1.19 Let p be a prime number, so D_{2p} is a dihedral group of order $2p$. Elements $a, b \in D$ are selected uniformly at random (with replacement). Thus each element is equally likely to be a, and similarly each element is equally likely to be b, and moreover it is perfectly possible that $a = b$. What is the probability that $D_{2p} = \langle a, b \rangle$? Note that you may have to analyze the case $p = 2$ separately.

1.20 Repeat the previous question, but with $m = p^t$ ($t \geq 2$) a power of the odd prime number p.

There are a vast and diverse collection of groups generated by an involution and an element of order 3, so this method of pushing back the frontiers of knowledge is in trouble. We can add extra restrictions. William Burnside (1852–1927) asked if a finitely generated group in which every element has finite order can be infinite. The affirmative answer was provided by E. S. Golod who constructed an infinite group of this type in 1964.

A group is said to have exponent $n > 0$ if the order of every element of G divides n. Novikov and Adian shown that there is an infinite group of exponent n on just two generators whenever $n \geq 4381$ and n is odd, and Adian later lowered the bound to $n \geq 665$.

Burnside asked a second more subtle question. If you specify the number n of generators of a finite group G, and insist that every element has finite order dividing m, is there a bound on the size of G? This question was resolved affirmatively when m is prime by Kostrikin. The positive answer for general m was obtained by Efim Zelmanov, an argument which won him a Fields medal in 1994. Fields medals are the mathematical equivalent of Nobel prizes. Zelmanov's proof is based on earlier work of Graham Higman and Philip Hall which reduced the problem to studying the case when m is a power of a prime number. The Hall–Higman reduction depends on the classification of very important objects called finite simple groups.

2
Structure

Consider the famous map from the real numbers to the positive real numbers $exp : \mathbb{R} \to \mathbb{R}^+$ defined by $r \mapsto e^r$ where e has suddenly (but temporarily) become the base of natural logarithms. The reals comprise a group under addition and the positive reals $\mathbb{R}^+$ form a group under multiplication. Our map, is bijective, and its inverse is $\ln : \mathbb{R}^+ \to \mathbb{R}$. These maps respect the group structures in the following sense:

$$\exp(x + y) = \exp(x)\exp(y) \; \forall x, y \in \mathbb{R}$$

and

$$\ln(ab) = \ln(a) + \ln(b) \; \forall a, b \in \mathbb{R}^+.$$

In each case, it does not matter whether you first combine the group elements, and then apply the map, or instead apply the map first, and then combine the group elements. It is no accident that exp and ln have this splendid property, and indeed it is one of many reasons why these particular functions play such an important role in mathematics. These maps are always written on the left for cultural reasons (i.e. for no reason at all).

Definition 2.1

Suppose that G and H are groups. A map $\varphi : G \to H$ is a *homomorphism* if whenever x, y in G, then $(xy)\varphi = (x)\varphi \cdot (y)\varphi$, or if writing maps on the left, $\varphi(xy) = \varphi(x) \cdot \varphi(y)$.

Thus a map between groups is a homomorphism if it preserves the multiplicative structure. Given any group G, there is a unique map $G \to 1$ to the trivial group which is certainly a homomorphism. Also if $H \leq G$ then the inclusion map inc $: H \to G$ sending each $h \in H$ to $h \in G$ is a homomorphism. It is a formality to check that a composite of homomorphisms is a homomorphism.

We now look at some more interesting examples. Let G be an abelian group. Fix any integer n, then the map $\psi_n : G \to G$ defined by $a \mapsto a^n$ for every $a \in G$ is a homomorphism. To see this, simply note that $(xy)^n = x^n y^n$ in any abelian group (even if n is negative!). In fact abelian groups are often written additively, and in that notation we would write $\psi_n : a \mapsto na$ for every $a \in G$.

The set of rational numbers $\mathbb{Q}$ forms a group under addition, and for $n \neq 0$ the homomorphism $\psi_n : a \mapsto na$ is a bijection in this case.

It is all very well for us to say that group homomorphisms are maps which preserve multiplicative structure, but there is more to a group than multiplication. Should we not insist that a homomorphism send 1 to 1 and also be compatible with inversion? Well, perhaps we should, but in fact there is no need to do so. The group axioms take care of everything, and this extra structure is respected automatically, as we now demonstrate.

Proposition 2.2

Suppose that $\varphi : G \to H$ is a homomorphism of groups. It follows that

(i) $(1_G)\varphi = 1_H$, and

(ii) if $x \in G$, then $(x^{-1})\varphi = ((x)\varphi)^{-1}$.

Proof

(i) Notice that $1_G^2 = 1_G$ so $(1_G 1_G)\varphi = (1_G)\varphi$. By the homomorphism property we have that $(1_G)\varphi \cdot (1_G)\varphi = (1_G)\varphi$. Now apply Lemma 1.1 (in H) to deduce that $(1_G)\varphi = 1_H$.

(ii) We have $1_H = (1_G)\varphi = (x^{-1}x)\varphi = (x^{-1})\varphi \cdot (x)\varphi$. By the uniqueness of inverses we conclude that $(x^{-1})\varphi = ((x)\varphi)^{-1}$.

$\square$

Thus homomorphisms respect local structure in that they do the right things with elements. They also interact well with subgroups, as the next proposition demonstrates.

Proposition 2.3

Suppose that $\varphi : G \to H$ is a homomorphism of groups where $A \leq G$ and $B \leq H$. Define two sets by

(i) $X = \{(g)\varphi \mid g \in A\}$ and

(ii) $Y = \{g \mid g \in G, (g)\varphi \in B\}$.

It follows that X is a subgroup of H and Y is a subgroup of G.

Proof

(i) $(1)\varphi = 1 \in X$ so $X \neq \emptyset$. Suppose that $x_1, x_2 \in X$, so $x_i = (g_i)\varphi$ for some $g_i \in A$ $(i = 1, 2)$. Now

$$x_1 x_2^{-1} = (g_1)\varphi \cdot (g_2^{-1})\varphi = (g_1 g_2^{-1})\varphi \in X$$

so $X \leq H$.

(ii) $(1)\varphi = 1 \in B$ so $1 \in Y \neq \emptyset$. Suppose that $y_1, y_2 \in Y$, so

$$(y_1 y_2^{-1})\varphi = (y_1)\varphi \cdot ((y_2)\varphi)^{-1} \in B$$

and therefore $Y \leq G$.

$\square$

Thus a homomorphism carries subgroups to subgroups, and the set of elements carried to a subgroup is a subgroup.

EXERCISES

2.1 Suppose that $\varphi : G \to L$ is a homomorphism of groups, and that $H \leq G$. Define a map $\varphi|_H : H \to L$ by $\varphi|_H : h \mapsto (h)\varphi$. Prove that $\varphi|_H$ is a group homomorphism. *Incidentally, $\varphi|_H$ is called the restriction of φ to H.*

2.2 Suppose that $\alpha : G \to H$ is a group homomorphism, $g \in G$ and $t \in \mathbb{Z}$. Prove that $(g^t)\alpha = ((g)\alpha)^t$.

2.3 Suppose that $\beta : G \to H$ is a group homomorphism, $x, y \in G$ and H is abelian. Prove that $(x^{-1}yx)\beta = (y)\beta$.

2.4 Suppose that $\gamma : G \to H$ is a surjective group homomorphism, and that G is an abelian group. Prove that H must be an abelian group.

2.5 Suppose that $\delta : G \to H$ is a group homomorphism, and that $g \in G$ has finite order n. Prove that $(g)\delta$ has finite order m for some m a divisor of n.

2.6 Suppose that $G = \langle X \rangle$ is a group generated by X. Suppose that $\theta_i : G \to H$ are group homomorphisms ($i = 1, 2$) such that $(x)\theta_1 = (x)\theta_2$ for all $x \in X$. Prove that $\theta_1 = \theta_2$.

2.7 Suppose that G is a group and that the map $\sigma : G \to G$ defined by $g \mapsto g^2$ for each $g \in G$ is a homomorphism. Prove that G is abelian.

2.1 Conjugacy

Suppose that G is a group and $x, y \in G$. We define x^y to be $y^{-1}xy \in G$. We say that x^y is a *conjugate* of x, and that y is an element which conjugates x to x^y. Notice that x and y commute if and only if $x^y = x$. It also follows that $x^y = x$ if and only if $y^x = y$. Also observe that $x^x = x$ for every $x \in G$, and that $1^x = 1$ for every $x \in G$.

Proposition 2.4

When $a, b, c \in G$ and $n \in \mathbb{Z}$ we have the following equations:

(i) $(ab)^c = (a^c)(b^c)$.

(ii) $(a^b)^c = a^{(bc)}$.

(iii) $(a^n)^b = (a^b)^n$.

These results show the joy of the x^y notation for $y^{-1}xy$. They have a casual familiarity evoking cosy schoolroom memories. In other words, the distracted or reckless user of this notation is likely to get the right answer (and you can ask no more of notation than it should be an effective substitute for thought). All you have to do is to forget that x^y means $y^{-1}xy$, and pretend that you are raising a variable to a power. However you must not get too confident, for if $a, b, c \in G$, there is no guarantee that $a^{bc} = a^{cb}$.

Proof

(i) $(ab)^c = c^{-1}abc = c^{-1}ac \cdot c^{-1}bc = (a^c)(b^c)$.

(ii) $(a^b)^c = c^{-1}(b^{-1}ab)c = (bc)^{-1}a(bc) = a^{bc}$.

(iii) If $n = 0$, then the result is clear. The result also holds when $n = 1$, so we can use part (i) and induction to deduce the result for all $n \in \mathbb{N}$. Now $(a^{-1})^b (a)^b = 1$ by part (i), so $(a^{-1})^b = (a^b)^{-1}$. Now use part (i) and induction again to deduce the result for all negative integers n.

$\square$

Lemma 2.5

Suppose that G is a group. For $a, b \in G$ we write $a \sim b$ if and only if there is $x \in G$ such that $a^x = b$. It follows that $\sim$ is an equivalence relation on G.

Proof

For every $a \in G$ we have $a^1 = 1 \cdot a \cdot 1 = a$ so reflexivity is established. Next suppose that $a, b \in G$ and that $a \sim b$, so there is $x \in G$ such that $a^x = b$, or rather $x^{-1}ax = b$. Premultiply by x and postmultiply by x^{-1} so that $a = xbx^{-1} = b^{(x^{-1})}$. Thus $b \sim a$.

Now for transitivity. Suppose that $a, b, c \in G$ and that both $a \sim b$ and $b \sim c$. Thus there are $x, y \in G$ with $a^x = b$ and $b^y = c$. Now

$$a^{xy} = (a^x)^y = b^y = c$$

and so $a \sim c$.

$\square$

Definition 2.6

The equivalence classes of the relation $\sim$ of Lemma 2.5 are called the *conjugacy classes* of G.

Notice that the conjugacy class containing 1_G is a singleton set. More generally the conjugacy class of $a \in G$ has size 1 if and only if $a^x = x^{-1}ax = a$ for every $x \in G$. In turn, this happens if and only if $xa = ax$ for every $x \in G$. Thus an element of G is in a conjugacy class of size 1 if and only if it commutes with each element of G.

It is instructive to look back to Example 1.15 to see a lot of conjugacy going on. In particular, if A and B are the matrices of the same linear transformation $S : V \to V$ with respect to rival bases (u_i) and (v_i), then A and B are conjugate in the group $\mathrm{GL}_n(k)$.

2.2 Normal Subgroups

Definition 2.7

Suppose that G is a group and that $N \leq G$. We say that N is a *normal subgroup* of G if whenever $g \in G$, then $gN = Ng$ (or equivalently $g^{-1}Ng = N$). We then write $N \trianglelefteq G$. If we wish to emphasise that a normal subgroup M is strictly included in G, then we leave out the line below the triangle and write $M \triangleleft G$.

Given that $g \in G$, and $N \trianglelefteq G$, we know that $gN = Ng$. In other words we have an equality of sets

$$\{gn \mid n \in N\} = \{ng \mid n \in N\},$$

but this does not necessarily mean that $gn = ng$ for each $n \in N$. The statement that $gN = Ng$ means that g commutes with N *as a set*, but this does not force g to commute with each individual element of N. However, if H is a subgroup of G and each element of H commutes with every element of G, then $H \trianglelefteq G$.

Proposition 2.8

Suppose that N is a subgroup of a group G. It follows that $N \trianglelefteq G$ if and only if $g^{-1}ng \in N$ for every $n \in N$ and $g \in G$.

Proof

If $N \trianglelefteq G$, then $gN = Ng$ for all $g \in G$. Thus $N = g^{-1}Ng$ so $g^{-1}ng \in N$ for every $n \in N$ and $g \in G$.

Now for the reverse implication. Suppose that $g^{-1}ng \in N$ for every $n \in N$ and $g \in G$. Therefore $g^{-1}Ng \subseteq N$. Replacing g by g^{-1} we have $gNg^{-1} \subseteq N$ and so $N = g^{-1}gNg^{-1}g \subseteq g^{-1}Ng$. We have both inclusions so $N = g^{-1}Ng$ for every $g \in G$. Thus $N \trianglelefteq G$.

$\square$

Definition 2.9

Suppose that G is a group. We define the centre $Z(G)$ of G by

$$Z(G) = \{x \mid x \in G, xy = yx \ \forall y \in G\}.$$

In words then, $Z(G)$ consists of those elements of G which commute with every element of G. It is routine to verify that $Z(G)$ is a subgroup of G, and that $Z(G) \trianglelefteq G$. The elements of $Z(G)$ are said to be *central* in G.

This notation for the centre (center in North America) is standard, and derives from the fact that the first letter of the word "centre" is Z (if you happen to be writing German).

Returning to the issue of normal subgroups, if $N \trianglelefteq G$ it does not necessarily follow that $N \leq Z(G)$. However, if $H \leq Z(G)$, then it is true that $H \trianglelefteq G$. In particular, if G is an abelian group, then all its subgroups are normal.

Remark 2.10

Observe that the subgroup N is normal in G if and only if $g^{-1}Ng = N$ for every $g \in G$. Thus if $n \in N$, then $n^g \in N$ for every $g \in G$, so N is a union of conjugacy classes of G. The argument is reversible, so if $H \leq G$ and H is a union of conjugacy classes of G, then $H \trianglelefteq G$.

Proposition 2.11

Suppose that $H \leq G$ and $N \trianglelefteq G$. It follows that $HN \leq G$.

Proof

We deploy Proposition 1.11 to show that $HN \leq G$. Now $1_H 1_N \in HN$ so $HN \neq \emptyset$. Next suppose that $h_1, h_2 \in H$ and $n_1, n_2 \in N$, so $h_1 n_1, h_2 n_2$ are generic elements of HN. Now

$$h_1 n_1 (h_2 n_2)^{-1} = h_1 n_1 n_2^{-1} h_2^{-1} = h_1 h_2^{-1} h_2 (n_1 n_2^{-1}) h_2^{-1}.$$

Now $h_1 h_2^{-1} \in H$ and $h_2 (n_1 n_2^{-1}) h_2^{-1}$ is a conjugate of an element of the normal subgroup N, and so itself is an element of N. Thus $h_1 n_1 (h_2 n_2)^{-1} \in HN$ so $HN \leq G$.

□

2.3 Factor Groups

Suppose that $N \trianglelefteq G$. The set of left cosets G/N and the set of right cosets $N\backslash G$ are the same because $gN = Ng$ for each $g \in G$. In this circumstance, it is usual to stick to the notation G/N. The elements of G/N are actually subsets of G. We have remarked that subsets of G can be multiplied together (in an associative way) to produce subsets of G. When you multiply cosets of a normal subgroup, there is magic in the air. Suppose that $x, y \in G$, then we have

$$
\begin{aligned}
(Nx)(Ny) &= N(x(Ny)) = N((xN)y) = N((Nx)y) \\
&= N(N(xy)) = (NN)(xy) = N(xy) = Nxy.
\end{aligned}
$$

We have put in the brackets to illustrate the role of associativity in this situation. The important terms are the ones on the ends, $(Nx)(Ny) = Nxy$. The product of cosets of N is a coset of N. Thus multiplication (of subsets of G) induces a closed associative binary operation on G/N. Notice that $N = N1$ acts as a two-sided identity, and that for every $g \in G$ we have

$$
NgNg^{-1} = Ngg^{-1} = N = Ng^{-1}g = Ng^{-1}Ng
$$

so we have the existence of two-sided inverse elements. Thus G/N is a group, and the inverse element of Ng is Ng^{-1}.

Definition 2.12

In the notation we have been using, the group G/N is called the *factor group* or *quotient group* of G by N. We refer to the process of constructing G/N from G and N as *factoring out N*.

It is also worth noting that if $N \leq H \leq G$ and $N \trianglelefteq G$, then $N \trianglelefteq H$. This is because if $gN = Ng$ for every $g \in G$, then certainly $hN = Nh$ for every $h \in H$. It also follows that H/N is a subgroup of G/N.

There is a natural map $\pi : G \to G/N$ defined by $g \mapsto Ng$ for every $g \in G$, so each element of G maps to the coset of N to which it belongs. Moreover, given any coset Nx it follows that $(x)\pi = Nx$, so π is surjective. Since G and G/N are both groups, and π is defined so effortlessly, there is hope that perhaps π is a homomorphism. It is! To see this, just observe that for every $x, y \in G$ we have $(xy)\pi = Nxy = NxNy = (x)\pi \cdot (y)\pi$. This map π is called the *natural projection* from G to G/N.

Suppose that $H \leq G$ and $N \trianglelefteq G$. The natural projection $\pi : G \to G/N$ is a homomorphism and so carries subgroups to subgroups. The image of H under π is $\{hN \mid h \in H\}$. However, this is the same as $\{hnN \mid h \in H, n \in N\}$ so the image of H under π is the same as the image under π of the subgroup HN. The advantage of the group HN is that it definitely contains N as a subgroup, whereas H need not do so. Thus in each case the image group is HN/N. It is now clear that if $H \leq T \leq HN$, then the image under π of T is HN/N.

Recall from Chapter 1 that one of the various pieces of notation for a cyclic group of order n is $\mathbb{Z}/n\mathbb{Z}$. This must have seemed rather peculiar to anyone who had not seen it before. However, now everything should be clear. The additive group $\mathbb{Z}$ is infinite cyclic. The subgroup $\langle n \rangle$ generated by $n \in \mathbb{Z}$ is $\{nz \mid z \in \mathbb{Z}\}$. Now $n\mathbb{Z}$ is sometimes used to denote this subgroup. Thus $\mathbb{Z}/n\mathbb{Z}$ is the set of (left or right) cosets of $n\mathbb{Z}$ in $\mathbb{Z}$. Now this subgroup $n\mathbb{Z}$ has index n in $\mathbb{Z}$ and the factor group $\mathbb{Z}/n\mathbb{Z}$ is an (additive) cyclic group of order n generated by $1 + n\mathbb{Z}$. Those people who use $\mathbb{Z}$ as the generic additive infinite cyclic group therefore find it very tempting to use $\mathbb{Z}/n\mathbb{Z}$ as the generic additive cyclic group of order n.

Definition 2.13

A homomorphism $\alpha : G \to H$ of groups is an *epimorphism* if α is surjective (onto). A homomorphism α is a *monomorphism* if α is injective (1–1). We say that α is an *isomorphism* when α is a bijective homomorphism (so it is simultaneously an epimorphism and a monomorphism). Finally we say that α is an *endomorphism* if $H = G$.

Observe that if N is a normal subgroup of a group G, then the natural projection $\pi : G \to G/N$ is surjective, so we have the option of adding colour to the language by also calling it the *natural epimorphism*.

The next result should come as no surprise, given that we interpret "isomorphism" as having the informal meaning *is structurally the same as*. If G_1, G_2 are isomorphic groups, then we write $G_1 \simeq G_2$.

Proposition 2.14

Suppose that A, B, C are groups and that $\alpha : A \to B$, $\beta : B \to C$ are isomorphisms. It follows that $\alpha^{-1} : B \to A$ is an isomorphism and that $\alpha \circ \beta : A \to C$ are both isomorphisms.

Proof

The map α^{-1} exists since α is a bijection. The fact that α^{-1} preserves structure may remind the reader of Example 1.15. Suppose that $b_1, b_2 \in B$. Thus there are unique $a_1, a_2 \in A$ such that $(a_1)\alpha = b_1$ and $(a_2)\alpha = b_2$. Now

$$(b_1 b_2)\alpha^{-1} = ((a_1)\alpha \cdot (a_2)\alpha)\,\alpha^{-1} = ((a_1 a_2)\alpha)\alpha^{-1} = a_1 a_2 = (b_1)\alpha^{-1} \cdot (b_2)\alpha^{-1}$$

as required, and we now know that the inverse of an isomorphism is an isomorphism.

Next we tackle composition of isomorphisms. The composite of bijections is a bijection, so the only issue is preservation of structure. Suppose that $a_1, a_2 \in A$. We have

$$(a_1 a_2)(\alpha \circ \beta) = ((a_1)\alpha \cdot (a_2)\alpha) \beta = (a_1)(\alpha \circ \beta) \cdot (a_2)(\alpha \circ \beta)$$

and so $\alpha \circ \beta$ is a homomorphism and thus an isomorphism.

$\square$

If there is an isomorphism between two groups, then we say that they are isomorphic. A pair of isomorphic groups must be structurally identical. Perhaps one is a group of matrices, and the other is a group of permutations. However, the isomorphism is such that it carries the (possibly infinite) multiplication table of one group to the multiplication table of the other. All group theoretic features of the groups must coincide. For example, they must have the same number of elements of order 2, and the same number of subgroups of infinite index. An isomorphism does more than match up the elements; it matches up the subgroups, their cosets, the conjugacy classes and so on.

EXERCISES

2.8 Suppose that G is a group and that $M, N \trianglelefteq G$. Prove that MN is a normal subgroup of G.

2.9 Suppose that G is a group and that $\{M_i \mid i \in I\}$ is a set of normal subgroups of G. Let $N = \cap_{i \in I} M_i$. Prove that $N \trianglelefteq G$. *Here the set I is an indexing set; it is there to provide labels (names) for the various groups M_i.*

2.10 Suppose that G is a group and that $H \leq G, N \trianglelefteq G$. Prove that $H \cap N \trianglelefteq H$.

2.11 Give an example of a group G with a subgroup H and an element $g \in G$ such that $H^g \subset H$ but $H^g \neq H$. *Hint: Consider 2×2 matrices with integer entries which have entry 0 in the bottom left position. Consider what happens to these matrices when they are conjugated by the 2×2 diagonal matrix*

$$\begin{pmatrix} 1/2 & 0 \\ 0 & 1 \end{pmatrix}.$$

2.12 Suppose that H is a subgroup of G, and that $|G : H| = 2$. Prove that $H \triangleleft G$.

2.13 Suppose that G is a group and that $H \leq Z(G)$, so all elements of H are central in G. Suppose furthermore that G/H is a cyclic group. Prove that G is abelian.

Proposition 2.15

Suppose that G is a group and that $X \subseteq G$. The following two subgroups coincide.

(a) The intersection of all normal subgroups N of G which have the property that $X \subseteq N$.

(b) The subgroup $\langle Y \rangle$ of G where $Y = \{x^g \mid g \in G\}$.

Proof

Let the group defined by (a) be N_1 and that defined by (b) be N_2. Notice that $Y \subseteq N$ whenever $X \subseteq N \trianglelefteq G$, so $Y \subseteq N_1$ and therefore $N_2 \leq N_1$.

On the other hand $X \subseteq Y$ so $X \subseteq N_2$. Also Y is invariant under conjugation so $N_2 \trianglelefteq G$. Therefore $N_1 \leq N_2$. We have both inclusions so equality is established.

$\square$

Let us celebrate by making a definition.

Definition 2.16

Using the notation established in Proposition 2.15, each of the groups defined by (a) and (b) is called the *normal closure* of X in G. This group is denoted $\langle X \rangle^G$. If $H \leq G$ we write H^G for the normal closure of H in G (rather than the correct but fussy $\langle H \rangle^G$).

Notice that if $M \trianglelefteq G$, then $M^G = M$.

2.4 Kernels and Images

Suppose that $\varphi : G \to H$ is a homomorphism of groups. This map gives rise to two interesting sets as follows:

(a) The image of φ is Im $\varphi = \{(g)\varphi \mid g \in G\}$.

(b) The kernel of φ is Ker $\varphi = \{k \mid k \in G, (k)\varphi = 1\}$.

Proposition 2.17

In the notation that we have just described, Im φ is a subgroup of H and Ker φ is a normal subgroup of G.

Proof

Everything save for the normality of Ker φ follows immediately from Proposition 2.3.

Suppose that $k \in$ Ker φ and $g \in G$. It follows that

$$(k^g)\varphi = (g^{-1}kg)\varphi = (g^{-1})\varphi \cdot 1_H \cdot (g)\varphi = (g^{-1}g)\varphi = (1_G)\varphi = 1_H.$$

Thus $k^g \in$ Ker φ, so Ker φ is a union of conjugacy classes, and so is a normal subgroup by Remark 2.10, or if you prefer by Proposition 2.8.

$\square$

In general it need not be the case that Im φ is a normal subgroup of H. Indeed, to witness this we can take the inclusion map of any non-normal subgroup of H into H. For example, H might be the symmetric group S_3 and we might be including one of its three non-normal subgroups of order 2.

When we have a homomorphism $\varphi : G \to H$ with $A \leq G$ and $B \leq H$, we may consider the image of the map φ when its domain is restricted to A. By Proposition 2.3 this set, sometimes written $(A)\varphi$, is a subgroup of H. On the other hand, we may consider the *preimage* of B. This is $\{g \mid g \in G, (g)\varphi \in B\}$. The preimage of B, sometimes written as $(B)\varphi^{-1}$, is a subgroup of G also by Proposition 2.3.

Proposition 2.18

Suppose that $\varphi : G \to H$ is a group homomorphism.

(i) φ is an epimorphism if and only if Im $\varphi = H$.

(ii) φ is a monomorphism if and only if Ker $\varphi = 1$.

Proof

Part (i) is a triviality, so we address part (ii). Let $K =$ Ker φ. Choose any $(x)\varphi \in$ Im φ. Let $\Omega_x = \{y \mid y \in G, (y)\varphi = (x)\varphi\}$, so certainly $x \in \Omega_x$. We now show that $Kx = \Omega_x$. Certainly if $k \in K$, then

$$(kx)\varphi = (k)\varphi \cdot (x)\varphi = 1_H \cdot (x)\varphi = (x)\varphi$$

so that $Kx \subseteq \Omega_x$. Conversely, if $y \in \Omega_x$, then $(y)\varphi = (x)\varphi$ so $(yx^{-1})\varphi = (y)\varphi((x)\varphi)^{-1} = 1$ and therefore $yx^{-1} \in K$. Thus $y \in Kx$, but y was an arbitrary element of Ω_x and so $\Omega_x \subseteq Kx$. We conclude that $\Omega_x = Kx$.

Now φ is injective if and only if each Ω_x is a singleton set, and this happens if and only if $K = 1$.

$\square$

We could equally well have shown that $\Omega_x = xK$ for every $x \in G$, so $xK = Kx$ for every $x \in G$. This yields an alternative proof that Ker $\varphi \trianglelefteq G$. Also note that the proof shows that if a group homomorphism fails to be injective, then it does so in a very uniform way, with the elements of G mapping to a given image point comprising a single coset of K. In the event that K happens to be finite, each coset of K has size $|K|$.

In the course of the proof we defined a set Ω_x to be the subset of G consisting of those elements which map to the same place as x. Notice that $(y)\varphi = (x)\varphi$ if and only if $\Omega_x = \Omega_y$, and this happens if and only if $Kx = Ky$. Some people use the following abuse of notation. If $y \in H$, then $(y)\varphi^{-1} = \{x \mid x \in G, \, (x)\varphi = y\}$. Note that $(y)\varphi^{-1} \neq \emptyset$ if and only if $y \in \text{Im } \varphi$. This notation is a little worrying, because φ is not necessarily injective, so there is not necessarily an inverse map φ^{-1}. The topologists call $(y)\varphi^{-1}$ the fibre of y (or the fiber of y if they are North American topologists). There is no further use for the Ω_x notation, and we consign it to history.

Choose and fix an epimorphism $\alpha : G \to H$ and let $N = \text{Ker } \alpha$. To each $L \leq G$ we associate $L^* = \{(l)\alpha \mid l \in L\}$. Since L^* is the image of a group under a homomorphism, it follows that $L^* \leq H$. Conversely, if $M \leq H$, then let $M^\dagger = \{x \mid x \in G, (x)\alpha \in M\} \leq G$. If $L \leq G$ then $(L^*)^\dagger = LN$, and if $M \leq H$, then $(M^\dagger)^* = M$. The situation becomes very clear if we restrict attention to those subgroups of G which contain N, as we see in the following result.

Theorem 2.19 (The Correspondence Principle)

We use the notation established in the preceding paragraph. The maps $*$ and $\dagger$ are a mutually inverse pair of bijections between the set of subgroups of G which contain N, and the subgroups of H. Suppose that $N \leq A, B \leq G$. These maps enjoy the following properties:

(i) $A \leq B$ if and only if $A^* \leq B^*$.

(ii) If $A \leq B$, then $|B : A| = |B^* : A^*|$.

(iii) $(A \cap B)^* = A^* \cap B^*$.

(iv) $\langle A, B \rangle^* = \langle A^*, B^* \rangle$.

(v) $A \trianglelefteq B$ if and only if $A^* \trianglelefteq B^*$.

(vi) A and B are conjugate in G if and only if A^* and B^* are conjugate in H.

Proof

If $L \leq G$ then

$$(L^*)^\dagger = \{g \mid g \in G, (g)\alpha = (l)\alpha \text{ for some } l \in L\}$$

$$= \{g \mid g \in G, (l^{-1}g) \in N \text{ for some } l \in L\} = LN.$$

In the event that $N \leq L$, it follows that $(L^*)^\dagger = L$. Now if $M \leq H$, then $(M^\dagger)^* = M$ is clear. Notice that $N \leq M^\dagger$. We have established that $*$ and $\dagger$ are mutually inverse bijections between the specified domains.

Parts (i) and (iv) are formalities and we hope that the reader can dispose of them in her head. We now deal with the remaining parts.

(ii) Suppose that $\{b_i \mid i \in I\}$ is a right transversal of A in B, so $B = \cup_i Ab_i$, and therefore $B^* = \cup_i (Ab_i)\alpha = \cup_i A^*(b_i)\alpha$. The only thing that would prevent $\{(b_i)\alpha \mid i \in I\}$ being a right transversal for A^* in B^* is $B^*(b_{i_1})\alpha = B^*(b_{i_2})\alpha$, but then $(b_{i_1}b_{i_2}^{-1})\alpha \in B^*$ so $b_{i_1}b_{i_2}^{-1} \in (B^*)^\dagger = B$ and so (happily) $b_{i_1} = b_{i_2}$.

(iii) The inclusion $(A \cap B)^* \subseteq A^* \cap B^*$ follows from part (i) because $A \cap B \subseteq A$ and $A \cap B \subseteq B$. Now suppose that $x \in A^* \cap B^*$. Since α is surjective there is $y \in G$ such that $(y)\alpha = x$, and $y \in A \cap B$ since $*\dagger = \text{id}$. Now $x = (y)\alpha \in (A \cap B)^*$. Thus $A^* \cap B^* \subseteq (A \cap B)^*$ and so finally $A^* \cap B^* = (A \cap B)^*$.

(v) If $A \trianglelefteq B$ then $A^b = A$ for all $b \in B$. Apply α so that $(A^*)^c = A^*$ for all $c \in B^*$. Conversely, suppose that $A^* \trianglelefteq B^*$. Now $(a)\alpha^{(b)\alpha} = (a^b)\alpha \in A^*$ for all $a \in A, b \in B$. Thus $A \trianglelefteq B$.

(vi) Suppose that A and B are conjugate. Thus there is $g \in G$ such that $A^g = B$, so $B = \{a^g \mid a \in A\}$. Apply α so $B^* = \{(a)\alpha^{(g)\alpha} \mid a \in A\} = (A^*)^{(g)\alpha}$. Thus A^* and B^* are conjugate.

Conversely, suppose that A^* and B^* are conjugate, so there is $g \in G$ such that $B^* = \{(a)\alpha^{(g)\alpha} \mid a \in A\}$. Taking preimages under α we see that $B = A^g$ and we are done.

$\square$

We illustrate this correspondence by returning to the group D of orientation-preserving symmetries of a cube introduced in Example 1.16. Recall that $|D| = 24$. Take three colours, called 1,2 and 3, and paint each pairs of opposite faces of the cube with the same colour, and use each colour on exactly two faces. Define an epimorphism $\zeta : D \to S_3$ as follows. If $g \in D$, then $(g)\zeta \in S_3$ has the property that $g)\zeta : i \mapsto j$ exactly when g sends the pair of faces coloured i to the positions previously occupied by the pair of faces coloured j, where $i, j \in \{1, 2, 3\}$. It is easy to see that $(g_1 g_2)\zeta = (g_1)\zeta \cdot (g_2)\zeta$ for all $g_1, g_2 \in D$. It is also clear from geometrical considerations that ζ is surjective. The kernel of ζ consists of the identity element, and rotations through π about axes of symmetry joining the centre of a face to the centre of the opposite face, a group of size 4 which is disguised version of the Vierergruppe mentioned in Chapter 1. To celebrate this observation we call this subgroup V.

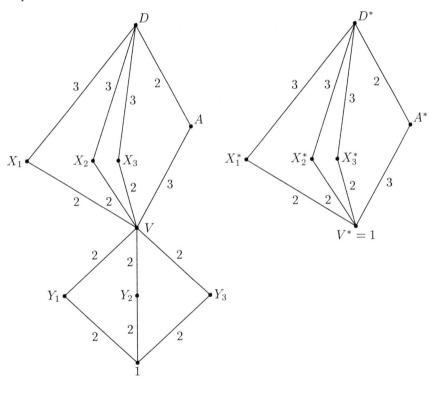

$$D \cong S_4 \qquad\qquad D^* \cong S_3$$

Figure 2.1 Correspondence between subgroups

Now, the full collection of subgroups of D is very complicated, but in Figure 2.1 we show the subgroups of D which happen either to contain or to be contained in V. We asserted in Chapter 1 that D is a copy of S_4, and we will justify this in Example 3.25.

As you can see, and should expect, the collection of subgroups of D which contain V intersect and join in a way which is also expressed by the subgroups of $D^* = S_3$. There are four groups strictly between V and D, and we will now animate them.

Pick any pair of opposite faces, and colour them both yellow. Now consider the collection of elements of D which preserve this colouring. There are 8 such elements; you can rotate about the axis through the centres of the coloured faces, or you can rotate the cube to exchange the coloured faces, and again spin about the axis through the centres of the coloured faces.

This group is morally the same as (i.e. isomorphic to) the dihedral group D_8 associated with the regular 4-gon (which we used to call a square until education got the better of us). There are three pairs of opposite faces we might choose to colour, and so three such dihedral subgroups X_1, X_2, X_3 of D arise.

To "see" the fourth group A between V and D, instead of colouring the faces we colour the vertices. Pick any vertex, and colour it black. Colour its neighbours white, their neighbours black and so on. See Figure 2.2.

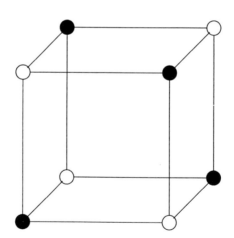

Figure 2.2 A decorated cube

Now, some elements of D preserve the colouring (for example, a rotation about a long diagonal), and some do not (for example, rotation through $\pi/2$

about an axis joining the centre of a face to the centre of the opposite face).
In fact there are 12 elements which preserve the colouring, and they form a
subgroup (of course). It is not really obvious that we have made an exhaustive
list of the subgroups strictly between V and D, unless of course you regard the
statement of Theorem 2.19 as obvious. When you do, everything is clear. It is
easy to verify that there are exactly four subgroups strictly between 1 and S_3,
and that is an end to the matter! Note that the subgroup of order 3 in S_3 is
normal in S_3, just as the subgroup A of order 12 is normal in D. Also, the three
subgroups of order 8 in D are conjugate, as are the three subgroups of order 2
in S_3. This is exactly as you would expect from Theorem 2.19.

2.5 Isomorphisms

In Chapter 1 we tried to persuade the reader that groups arise naturally when
considering a set S equipped with structure, and then studying those bijections
from X to X which preserve the structure, and whose inverses also preserve
the structure. It does not actually matter what the structure is; it could be an
ordering, some sort of geometry, a distance function or almost any algebraic
confection. Closing the circle, we can even fix a particular group G and consider
the collection of all isomorphisms from G to itself. Such self-isomorphisms are
called *automorphisms* of G. We know that the inverse of a group isomorphism
is a group isomorphism, and it is a formality to check that Aut G, the set of all
self-isomorphisms from G to G, is a group under the operation "composition
of maps".

Note that Id_G is the identity element of this group. Consider the map τ_g :
$G \to G$ defined via cnjugation by some $g \in G$. Thus

$$\tau_g : G \to G, \quad (x)\tau_g = g^{-1}xg \ \forall x \in G.$$

First observe that τ_g is a bijection, since for all $x \in G$ we have $x =
gg^{-1}xgg^{-1} = g^{-1}gxg^{-1}g$ and so

$$\tau_g \cdot \tau_{g^{-1}} = \tau_{g^{-1}} \cdot \tau_g = \mathrm{Id}_G. \tag{2.1}$$

We are using the fancy characterization of bijections in terms of mutually
inverse maps. Next we show that each τ_g is a homomorphism; suppose that
$x, y \in G$ then

$$(xy)\tau_g = g^{-1}xyg = g^{-1}xg \cdot g^{-1}yg = (x)\tau_g \cdot (y)\tau_g$$

and each τ_g is a bijective homomorphism with inverse $\tau_{g^{-1}}$. Thus each τ_g is an
automorphism of G. These automorphisms τ_g are called *inner automorphisms*
of G.

If $g \notin Z(G)$, then there is $x \in G$ such that $(x)\tau_g \neq x$. Thus τ_g is not the identity map. Note that if $g, h \in G$, then $\tau_g \cdot \tau_h = \tau_{gh}$. This is because when we apply either map to $x \in G$ we get $h^{-1}g^{-1}xgh$. This important equation tells us that we have a homomorphism

$$\tau : G \to \text{Aut } G \text{ defined by } x \mapsto \tau_x.$$

Now Im τ is therefore a subgroup denoted Inn G of Aut G. Its elements are called *inner automorphisms* of G, and the group itself is called the *inner automorphism group* of G. Notice that Ker $\tau = Z(G)$, the centre of G.

These maps τ_x ensure that every non-abelian group has a non-trivial automorphism. If G is an abelian group, each τ_g is the identity map (the trivial automorphism). In this event however, the map inv $: G \to G$ defined by $x \to x^{-1}$ for every $x \in G$ is an automorphism since $u^{-1}v^{-1} = v^{-1}u^{-1} = (uv)^{-1}$ for every $u, v \in G$. This will be a non-trivial automorphism unless every element of G has order dividing 2. We can then write this group additively, and view it as a vector space over the field[1] with 2 elements. The reader who knows enough linear algebra should try to complete the argument that shows that the only groups G with the property that Aut G is trivial are the trivial group and the cyclic group of order 2. It helps to know (or believe) that every vector space has a basis, otherwise you may be in trouble. See Theorem 8.9. The point is that if G is an infinite group, the vector space we have just built will not be finite dimensional.

An automorphism of a group G will send subgroups to subgroups, and normal subgroups to normal subgroups. In fact we can express normality in terms of automorphisms. A subgroup N of G is normal if and only if it is invariant under all inner automorphisms (conjugations by elements of G). We can strengthen this condition as follows.

Definition 2.20

We say that H is a *characteristic subgroup* of G if every automorphism of G induces an automorphism of H.

Definition 2.21

We say that H is a *fully invariant* subgroup of G if every endomorphism of G induces an endomorphism of H. We write H f.i. G.

Thus H f.i. G implies H char G and H char G implies $H \trianglelefteq G$.

---

[1] Appendix A on Fields is supplied

Proposition 2.22

Suppose that G is a group, and that $H \leq K \leq G$.

(i) If H char K and K char G, then H char G.

(ii) If H f.i. K and K f.i. G, then H f.i. G.

(iii) If H char K and $K \trianglelefteq G$, then $H \trianglelefteq G$.

Proof

(i) Suppose that $\alpha \in$ Aut G. Now α with its domain restricted to K is $\beta \in$ Aut K. Next notice that H char K, so if we restrict the domain of β to H we obtain $\gamma \in$ Aut H. Now for every $h \in H$ we have $(h)\gamma = (h)\beta = (h)\alpha$, so $\gamma \in$ Aut H is obtained by restricting the domain of α to H. It now follows that H char G.

(ii) The proof of this result is analogous to the proof of part (i). The only difference is that we replace Aut X by End X (the set of endomorphisms of X) where appropriate.

(iii) Once more, the structure of the proof repeats the previous pattern. Suppose that $g \in G$, and let conjugation by g be effected by the automorphism τ_g of G. Since $K \trianglelefteq G$, the map τ_g induces $\beta \in$ Aut K by restriction of domain ($\tau_{g^{-1}}$ induces the inverse of β so that is why β is bijective). Now H char K, so if we restrict the domain of β to H we obtain $\gamma \in$ Aut H. Now for every $h \in H$ we have $(h)\gamma = (h)\beta = (h)\alpha$, so $\gamma \in$ Aut H. However, $(h)\gamma = (h)\tau_g = g^{-1}hg$ for every $h \in H$ and $g \in G$. Therefore $g^{-1}Hg = H$ for every $g \in G$, so $H \trianglelefteq G$.

$\square$

Remark 2.23

We trail a notational technique which we will deploy in Chapter 5. When thinking about G and Aut G simultaneously, it can be rather confusing to write the elements of these two groups in the same line of print. The two groups become visually scrambled as the symbols intermix, and that does not help us to reason about what is happening. A useful trick is to write x^α as an alternative notation for $(x)\alpha$ when $x \in G$ and $\alpha \in$ Aut G. If $\alpha = \tau_g$ is an inner automorphism (conjugation by g) then we now have four equally correct ways of writing the value of τ_g applied to x. They are

$$(x)\tau_g, x^{\tau_g}, g^{-1}xg \text{ and } x^g.$$

We can use $x^{-\alpha}$ instead of $(x^{-1})^\alpha$ which is the same as $(x^\alpha)^{-1}$.

From a practical point of view, it is important not to swap between notations too frequently in the course of an argument so as to give the reader a sporting chance of understanding what is going on. However, the ability to choose notation is a valuable asset, because you are free to select the mode of expression which most appropriately suits the context. The expression $(x)\alpha$ is very clear as a piece of isolated notation, but x^α is much more useful when building a large complicated expression involving the application of automorphisms to elements of G, multiplications and inversions. For example, $(x)\alpha((x)\alpha)\beta((x^{-1})\beta)\alpha$ compares unfavourably with $x^\alpha x^{\alpha\beta} x^{-\beta\alpha}$. We can even stretch to using $+$ in the exponent, so that the last expression can be written $x^{\alpha+\alpha\beta-\beta\alpha}$.

Suppose that $H \leq G$. Just as we write $H = H^g$ to indicate equality between the sets H and $\{g^{-1}Hg \mid h \in H\}$, we can take $\alpha \in \operatorname{Aut} G$ and write $H = H^\alpha$ to indicate that $H = \{h^\alpha \mid h \in H\}$. Now we have a slick way of explaining when a subgroup H is characteristic; we have H char G if and only if $H = H^\alpha$ for every $\alpha \in \operatorname{Aut} G$. On the other hand $H \trianglelefteq G$ if and only of $H = H^\theta$ for all $\theta \in \operatorname{Inn} G$.

Informally, subgroups H of G can often be moved about by the application of automorphisms of G. A normal subgroup N is harder to shift; conjugation by each element of G keeps N in the same place as a set but may move the elements of N among themselves. A central subgroup K is a normal subgroup of G with the additional property that conjugation does not even move the elements of K among themselves. A characteristic subgroup L is one which cannot be moved as a set by any automorphism of G, though it may be that such an automorphism will move the elements of L among themselves.

Isomorphisms are one of the key notions in group theory (and in modern algebra in general). Any observation about isomorphisms is therefore likely to have many applications. The following theorems on isomorphisms make life so much simpler once you understand them. They enable you to take countless mathematical short-cuts. Therefore you are strongly recommended to try to incorporate them into your way of thinking about group theory, life and so on.

Theorem 2.24 (First Isomorphism Theorem)

Suppose that $\alpha : G \to H$ is a group epimorphism. Let $K = \operatorname{Ker} \alpha \trianglelefteq G$, and form the factor group G/K. We then have the natural epimorphism $\pi : G \to G/K$. There is a unique isomorphism $\widehat{\alpha} : G/K \to H$ such that $\alpha = \pi \circ \widehat{\alpha}$.

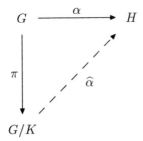

Figure 2.3 The first isomorphism theorem

Proof

We are required to show that the isomorphism indicated by the dotted line exists, and that the diagram *commutes* in the sense that if you compose maps as you run round the triangle, it does not matter which way you go. Provided you head in the direction of the arrows, the map you eventually produce will be independent of your route. That is a another way of saying that $\alpha = \pi \circ \widehat{\alpha}$.

First we define $\widehat{\alpha} : G/K \to H$ by $Kx \to (x)\alpha$ for every $x \in G$. The issue of well-definedness arises. The problem is that the coset Kx will have other descriptions (except in the very special case $|K| = 1$). Choosing any $y \in Kx$ we have $Kx = Ky$. Our definition will be nonsensical unless for every such y we have $(x)\alpha = (y)\alpha$. Happily for group theory, this is exactly what happens, for if $Kx = Ky$, then $xy^{-1} \in K$ so $(xy^{-1})\alpha = 1$ and thus $(x)\alpha((y)\alpha)^{-1} = 1$. We conclude that $(x)\alpha = (y)\alpha$ so our definition of $\widehat{\alpha}$ is legitimate. The rest of the proof will be downhill.

We next show that $\widehat{\alpha}$ is a homomorphism. Suppose that $Kx, Ky \in G/K$, so

$$(KxKy)\widehat{\alpha} = (Kxy)\widehat{\alpha} = (xy)\alpha = (x)\alpha \cdot (y)\alpha = (Kx)\widehat{\alpha} \cdot (Ky)\widehat{\alpha}.$$

This settles the issue. Also $\widehat{\alpha}$ is surjective by design. Suppose that $(Kx)\widehat{\alpha} = 1$ so $(x)\alpha = 1$. We conclude that $x \in K$ so $Kx = K = 1_{G/K}$. The kernel of $\widehat{\alpha}$ is therefore the trivial group. It consists of $\{1_{G/K}\}$ (which happens to be $\{K\}$). Proposition 2.18(ii) allows us to conclude that $\widehat{\alpha}$ is injective. We deduce that $\widehat{\alpha}$ is an isomorphism.

Now suppose that ρ is any rival map $\rho : G/K \to H$ such that $\pi \circ \rho = \alpha$. Now apply both maps to $g \in G$ so $(g)(\pi \circ \rho) = (g)\alpha$ and thus $(Kg)\rho = (g)\alpha = (Kg)\widehat{\alpha}$. Thus $\rho = \widehat{\alpha}$.

$\square$

We may still apply this theorem even if $\alpha : G \to H$ is not surjective, because we may discard the elements of H which are not in Im α so that we can think of $\alpha : G \to \mathrm{Im}\ \alpha$ (we should really change the name of α here since it has a new codomain, but the change is only cosmetic, so if you don't tell, we won't!). The first isomorphism applies with $\mathrm{Im}\ \alpha$ replacing H, so $G/\mathrm{Ker}\ \alpha \simeq \mathrm{Im}\ \alpha$ via a unique isomorphism $\widehat{\alpha}$ such that $\pi \circ \widehat{\alpha} = \alpha$.

This theorem has substantial philosophical meaning. It says that if $G \to H$ is an epimorphism of groups with kernel K, then a copy of H can be manufactured directly from G via the factor group G/K. In particular, if G happens to be finite, it has only finitely many subsets and therefore only finitely many normal subgroups. Thus there are only finitely many epimorphic images of G (up to isomorphism). The parenthetic remark is to indicate that we do not count isomorphic groups as being different for this purpose.

We began this section by observing that $\tau : G \to \mathrm{Aut}\ G$ defined by $g \mapsto \tau_g$ is a homomorphism with image $\mathrm{Inn}\ G$ and kernel $Z(G)$. The first isomorphism theorem now tells us that $G/Z(G) \simeq \mathrm{Inn}\ G$. In particular, if $Z(G) = 1$, then $G \simeq G/Z(G) \simeq \mathrm{Inn}\ G$ via the natural map $g \mapsto \tau_g$. Thus a group with a trivial centre *embeds* in its own group of automorphisms. When we say "embeds" we mean that there is an isomorphic copy of G occurring as a subgroup of $\mathrm{Aut}\ G$.

Example 2.25

Suppose that $G \simeq S_4$. It is easy to check that S_4 has a trivial centre and so $\mathrm{Aut}\ G$ has a subgroup isomorphic to S_4. In fact $\mathrm{Aut}\ G \simeq S_4$, but this is not necessarily obvious.

Every group G has the normal subgroups $1 \trianglelefteq G$ and $G \trianglelefteq G$, and these normal subgroups are distinct unless G is the trivial group. We shall say that these ubiquitous subgroups are improper normal subgroups. They arise as kernels of homomorphisms via the identity map $G \to G$ which has 1 as its kernel, and the (unique) map $G \to 1$ which has G as its kernel. It is particularly interesting when there are as few normal subgroups as possible. Motivated by this thought, we make an important definition.

Definition 2.26

A group $G \neq 1$ is a *simple* group if whenever $N \trianglelefteq G$, then $N = 1$ or $N = G$.

The abelian simple groups are quite straightforward.

Proposition 2.27

Let G be an abelian simple group. It follows that G is cyclic of prime order.

Proof

Since $G \neq 1$ there is $g \in G$ with $g \neq 1$. Now $1 \neq \langle g \rangle \trianglelefteq G$. All subgroups of G are normal since G is abelian. The simplicity of G forces $G = \langle g \rangle$. Now $n = o(g)$ is finite else $\langle g^2 \rangle$ is a non-trivial proper normal subgroup. Choose p a prime divisor of n then $\langle g^{n/p} \rangle \neq G$ and so it must be trivial. Therefore $n = p = |G|$.

$\square$

The situation is considerably more complicated for non-abelian simple groups. Our understanding of infinite simple groups is rather poor, and there seems no hope of classifying them. However, the finite simple groups have very probably been classified. There is a list of them, and that list is widely believed to be correct. However, the "proof" of the correctness of the list is so large (thousands of pages scattered across journals and desks) and was constructed by so many mathematicians (some of whom seem to have been a little ambitious with their claims) that there is slight room to doubt. This is a very unusual situation in mathematics. Most proofs are fairly short, and if the result is important, are checked by lots of very competent people. No human knowledge is ever certain of course, and it may be that everyone reading a proof is making the same mistake as its author. However, such knowledge is as sure as we are ever likely to get. The philosophical problem is not solved by making a proof sufficiently formal that it can be checked by a computer, since then we have to worry if there is a flaw in the computer program or in the hardware of which the machine is built.

As we write this book in 1999, a revisionist process is under way. A group of mathematicians is trying to organize the classification of finite simple groups into a coherent and accessible form, and is publishing their work as a series of books. If the text in your hands is fairly grubby, it may be that enough time has passed for this important enterprise to have succeeded (or perhaps the revisionists will have found an irreparable hole in the proof!). See the web site of this book for the latest information.

Theorem 2.28 (Second Isomorphism Theorem)

Suppose that $N \trianglelefteq G$ and $H \leq G$. It follows that HN is a subgroup of G, and moreover $H, N \trianglelefteq HN$. The composite α of the inclusion map $H \to G$ and the natural projection $\pi : G \to G/N$ has kernel $H \cap N$ and image HN/N. It follows that $(H \cap N)h \mapsto Nh$ is an isomorphism $H/H \cap N \simeq HN/N$.

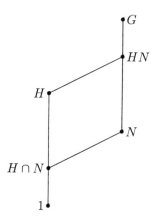

Figure 2.4 The second isomorphism theorem

Proof

We begin with the formalities. Since $N \trianglelefteq G$ we know that HN is a subgroup of G by Proposition 2.11. Now $H = H1 \subseteq HN$ and $N = 1N \subseteq HN$ and we have confirmed these inclusions. Note that $(h)\alpha = 1$ if and only if $(h)\pi = 1$, which happens if and only if $h \in N$. Thus $\mathrm{Ker}\, \alpha = H \cap N$ and we obtain the bonus information that $H \cap N \trianglelefteq H$ which will not shock the diligent reader who does all the exercises.

$$\mathrm{Im}\, \alpha = \{(h)\pi \mid h \in H\} = \{Nh \mid h \in H\} = HN/N.$$

We now apply the first isomorphism theorem so $H/\mathrm{Ker}\, \alpha \simeq \mathrm{Im}\, \alpha$, and a little bookkeeping shows that the map is as indicated.

$\square$

Theorem 2.29 (Third Isomorphism Theorem, memorable form)

Suppose that $N \leq K \leq G$ and that both N and K are normal subgroups of G. It follows that $K/N \trianglelefteq G/N$ and $(G/N)/(K/N) \simeq G/K$. Moreover, the isomorphism is given by the recipe $(K/N)Ng \mapsto Kg$.

Note that this version is memorable since a sufficiently blasé reader should be prepared simply to *cancel the N*.

Proof

Suppose that $Kg_1 = Kg_2$ for some $g_1, g_2 \in G$. Premultiply by N so $Ng_1 = Ng_2$. This shows that the map $\theta : G/K \to G/N$ such that $Kg \mapsto Ng$ is well-defined. Next we show that θ is a group homomorphism. Now for all $x, y \in G$ we have

$$(KxKy)\theta = (Kxy)\theta = Nxy = NxNy = (Kx)\theta(Ky)\theta.$$

Thus θ is a group homomorphism. It is clear that θ is surjective, and Ker $\theta = \{Kx \mid x \in N\} = N/K$. Now the first isomorphism theorem takes us home.

$\square$

Something can be salvaged from the wreckage even if we stop concentrating on groups, and just worry about collections of cosets.

Suppose that $H, K \leq G$, then HK is a union of right cosets of H in G. We call this collection of right cosets $H \backslash HK$ (even though HK need not be a group).

Proposition 2.30

There is a natural bijective correspondence between $H \backslash HK$ and $(H \cap K) \backslash K$.

Proof

Define a map $\gamma : H \backslash HK \to (H \cap K) \backslash K$ by $\gamma : Hk \to (H \cap K)k$. We first show that γ is well-defined. Suppose that $Hk_1 = Hk_2$ for $k_1, k_2 \in K$, then $k_1 k_2^{-1} \in H \cap K$ and so $(H \cap K)k_1 = (H \cap K)k_2$ and γ is well-defined. We show that γ is bijective by exhibiting δ, a two-sided inverse for γ. We define δ by $\delta : L \to HL$ where $L = (H \cap K)z \in (H \cap K) \backslash K$. We leave the formality of checking that γ and δ are mutually inverse to the interested reader.

$\square$

Corollary 2.31

If H and K are finite subgroups of G, then $|HK|/|H| = |K|/|H \cap K|$. This result assumes great importance when it is written symmetrically as

$$|HK| = \frac{|H||K|}{|H \cap K|}.$$

We have seen this before. Thus if G is a finite group with subgroups H, K with $|H|, |K| > \sqrt{|G|}$ then $H \cap K \neq 1$. Of course, you can squeeze out much more than that. The importance of this result is that the intersection of two "large" subgroups of a finite group must be "substantial". Now, the memorable form of the third isomorphism theorem is easily recalled by anyone prepared to *cancel* N, but really this version of the theorem is of limited utility. Instead, we recast it in a way which exposes its close relationship with the first isomorphism theorem.

Theorem 2.32 (Third Isomorphism Theorem, Useful Form)

Suppose we have a group epimorphism $\alpha : G \to H$ and $N \lhd G$ with $N \subseteq K = \text{Ker } \alpha$. We form the factor group G/N. We then have a natural projection $\pi : G \to G/N$. There is a unique homomorphism $\widehat{\alpha} : G/N \to H$ such that $\alpha = \pi \circ \widehat{\alpha}$, and $\text{Ker } \widehat{\alpha} = K/N$.

Proof

The proof is much the same as the proof we have given of the first isomorphism theorem. The map $\widehat{\alpha} : G/N \to H$ is defined by $Nx \mapsto (x)\alpha$. Thus $K/N = \text{Ker } \widehat{\alpha}$ and so $K/N \lhd G/N$.

$\square$

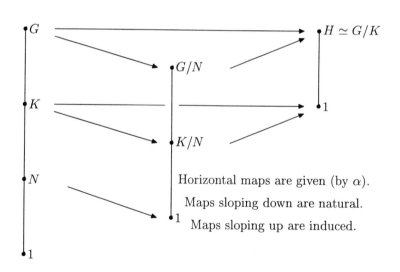

The key point is that the "top" triangle of maps commutes.

Figure 2.5 The third isomorphism theorem (useful form)

Incidentally, following the last proof we may apply the first isomorphism theorem to deduce that $(G/N)/(K/N) \simeq H \simeq G/K$, the memorable form of the third isomorphism theorem. Conversely, we can take the memorable form of the theorem and deduce the useful form with a little work. The reader is invited to do this.

Perhaps now is as good a time as any to introduce the *modular law*. Suppose that $A, B, C \leq G$. An optimist might hope that $A(B \cap C) = AB \cap AC$, but this is generally false. However, it is true provided A is a subgroup of either B or C.

Proposition 2.33 (The Modular Law)

Suppose that $A, B, C \leq G$ and that $A \leq B$, then

(i)

$$A(B \cap C) = AB \cap AC = B \cap AC.$$

(ii)

$$(B \cap C)A = BA \cap CA = B \cap CA.$$

Proof

We address part (i). Suppose that $a \in A$ and $x \in B \cap C$. Thus $ax \in AB$ and $ax \in AC$ so $A(B \cap C) \subseteq AB \cap AC$. Next suppose that $a_1 b = a_2 c$ for $a_1, a_2 \in A, b \in B$ and $c \in C$. Thus $a_1 b = a_2 c \in AB \cap AC$. Now $c = a_2^{-1} a_1 b \in B \cap C$ since $A \leq B$. Thus $a_2 c \in A(B \cap C)$. We have established that $A(B \cap C) = AB \cap AC$, and the fact that $A \leq B$ allows us to complete the proof of part (i). The proof of part (ii) is entirely similar and is omitted.

$\square$

Now for a beautiful result, widely known as Zassenhaus's Butterfly Lemma. You need to apply a liberal imagination to the Hasse diagram in Figure 2.6 in order to see why.

Lemma 2.34 (Zassenhaus)

Suppose that G is a group with subgroups A, B, C, D such that $A \trianglelefteq B$ and $C \trianglelefteq D$. It follows that $A(B \cap C) \trianglelefteq A(B \cap D)$ and $C(D \cap A) \trianglelefteq C(D \cap B)$, and moreover that

$$\frac{A(D \cap B)}{A(C \cap B)} \simeq \frac{C(B \cap D)}{C(A \cap D)}.$$

Remark 2.35

This remark is even more whimsical than the usual ramblings. Part of the elegance of this result comes from the symmetry. The binocular reader is strongly urged to keep one eye fixed on Figure 2.6 while reading the following with the other. If you swap A with C and B with D, the result remains the same. One way to look at this result is to think of how the fact that $C \trianglelefteq D$ manifests itself in other parts of the group. Below B, the manifestation is that $C \cap B \trianglelefteq D \cap B$. Now if we next focus on groups which are between A and B, the manifestation is that $A(C \cap B) \trianglelefteq A(D \cap B)$. If we start again, and see how $A \trianglelefteq B$ manifests itself elsewhere, we find that when we look below D we find $A \cap D \trianglelefteq B \cap D$ and that between C and D the manifestation is that $C(A \cap D) \trianglelefteq C(B \cap D)$. Thus $A \trianglelefteq B$ leaves a shadow between C and D, and $C \trianglelefteq D$ leaves a shadow between A and B. In some sense, Zassenhaus's Lemma says that the shadows are the same. As well as being pretty, this result will have spectacular application in Chapter 5.

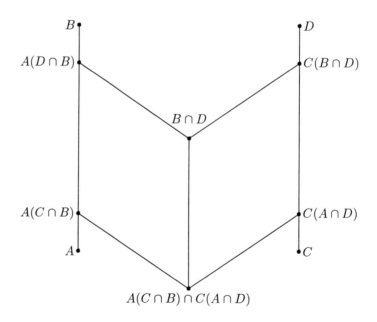

Figure 2.6 Zassenhaus's Butterfly Lemma

Proof

First we observe that the inclusion map $B \cap D \to D$ composed with the natural projection $D \to D/C$ has kernel $B \cap D \cap C = B \cap C$. It follows that $B \cap C \trianglelefteq B \cap D$. Now the natural projection $B \to B/A$ preserves normality, so $A(B \cap C) \trianglelefteq A(B \cap D)$. By symmetry, we also have that $C(A \cap D) \trianglelefteq C(B \cap D)$.

Now we go to work.

$$\frac{A(B \cap D)}{A(B \cap D)} = \frac{A(B \cap C)(B \cap D)}{A(B \cap C)} \text{ since } B \cap C \subseteq B \cap D.$$

We apply the second isomorphism theorem so

$$\frac{A(B \cap C)(B \cap D)}{A(B \cap C)} \simeq \frac{B \cap D}{B \cap D \cap A(B \cap C)}.$$

We make two observations about the final term in the "denominator" of this factor group. The first point is that A is normal in B, so $A(B \cap C) = (B \cap C)A$, and the second is that we can apply the modular law to each side of this equation ($A \subseteq B$) so each side is $B \cap AC = B \cap CA = B \cap AC \cap CA$. Similarly $C(A \cap D) = (A \cap D)C = D \cap AC \cap CA$ so

$$B \cap D \cap A(B \cap C) = B \cap D \cap CA \cap AC = A(B \cap C) \cap C(A \cap D).$$

We conclude that

$$\frac{A(B \cap D)}{A(B \cap C)} \simeq \frac{B \cap D}{A(B \cap C) \cap C(A \cap D)}.$$

Now, viewing the Hasse diagram in Figure 2.6, we see that arguing with the right wing of the butterfly instead of the left, we are done.

□

2.6 Internal Direct Products

Suppose that G is a group containing subgroups A and B. It may be that $AB = G$, so every $g \in G$ factorizes as $g = ab$ for $a \in A$ and $b \in B$. This expression will not be unique unless $A \cap B = 1$. We spell out the details. If $a_1 b_1 = a_2 b_2$ are rival expressions of a product, then $a_2^{-1} a_1 = b_2 b_1^{-1} \in A \cap B$. If $A \cap B = 1$, then this forces $a_1 = a_2$ and $b_1 = b_2$. On the other hand, if $A \cap B \neq 1$, then there is $1 \neq c \in A \cap B$, and 1 has the distinct rival factorizations 11 and cc^{-1}. The situation gets particularly interesting when both A and B are normal subgroups of G.

Definition 2.36

Suppose that the group G has normal subgroups A, B such that $AB = G$ and $A \cap B = 1$. We say that G is the (internal) direct product of A and B. We write $G = A \times B$.

In these circumstances, every element of G is uniquely expressible as ab for $a \in A, b \in B$.

Next we show that each element $a \in A$ commutes with each element $b \in B$. Where on earth does this come from? There is definitely something up the authorial sleeve and it is $a^{-1}b^{-1}ab$. This element leads a treble life, because

$$a^{-1}b^{-1}ab = (b^{-1})^a b = a^{-1}a^b.$$

Today's brilliant trick is tomorrow's standard technique. This particular piece of cunning will soon become Chapter 5. Elements of the shape $a^{-1}b^{-1}ab$ are called *commutators*.

Now $B \trianglelefteq G$ so $(b^{-1})^a \in B$ and therefore $(b^{-1})^a b \in B$. On the other hand, $A \trianglelefteq G$ so $a^b \in A$ and therefore $a^{-1}a^b \in A$. Thus $a^{-1}b^{-1}ab \in A \cap B = 1$. We deduce that $a^{-1}b^{-1}ab = 1$ and so $ab = ba$. The elements $a \in A$ and $b \in B$ were arbitrary, so this paragraph has accomplished its task.

Now we see that the multiplication is well adapted to the factorization: $(a_1 b_1)(a_2 b_2) = (a_1 a_2)(b_1 b_2)$. The uniqueness of factorization gives us two maps $\alpha : G \to A$ and $\beta : G \to B$ defined by $(ab)\alpha = a$ and $(ab)\beta = b$. One may verify that α, β are epimorphisms, and $\operatorname{Ker} \alpha = B$, $\operatorname{Ker} \beta = A$. We also have two inclusion maps $\operatorname{inc}_A : A \to G$ and $\operatorname{inc}_B : B \to G$. Note that $\operatorname{inc}_A \circ \alpha = \operatorname{Id}_A$ and $\operatorname{inc}_B \circ \beta = \operatorname{Id}_B$ where $\operatorname{Id}_A, \operatorname{Id}_B$ are the identity maps on A and B respectively.

Suppose instead that we are given groups A and B, and we seek a group G such that G is the internal direct product of A and B. We cannot quite manage to construct such a G, but we almost can.

Let $G = A \times B$ be the set of ordered pairs (a, b) with $a \in A, b \in B$. We endow this set with a binary operation in order to turn it into a group. Suppose that $(a_1, b_1), (a_2, b_2) \in A \times B$. We define $(a_1, b_1) * (a_2, b_2) = (a_1 a_2, b_1 b_2) \in A \times B$. This is clearly a binary operation on $A \times B$, and the associative law follows immediately from the associativity of the products on A and B. The two-sided identity element is $(1_A, 1_B)$ and the two-sided inverse of (a, b) is (a^{-1}, b^{-1}). Thus we have turned $A \times B$ into a group, called the external direct product of A and B. The biggest drawback of this group is that it does not have A and B as subgroups, but it almost does. Let $\widehat{A} = \{(a, 1) \mid a \in A\}$ and $\widehat{B} = \{(1, b) \mid b \in B\}$. The maps $a \mapsto (a, 1)$ for all $a \in A$ and $b \mapsto (1, b)$ for all $b \in B$ are isomorphisms from A and B to subgroups of G called $\widehat{A}$ and $\widehat{B}$ respectively. Notice that G is the internal direct product of $\widehat{A}$ and $\widehat{B}$, so G is only a whisper short of being the internal direct product of A and B. We write $G = A \times B$ as if the product were an internal one. We have to hope that the context makes it clear whether the direct product is internal or external.

In the event that we are working with abelian groups, and happen to be using additive notation, it is customary to write $A \oplus B$ instead of $A \times B$ and to refer to a direct sum rather than a direct product. Another name for a direct product is a *Cartesian* product.

We can generalize all this. Suppose that $G_1, \ldots, G_n$ is a list of finitely many groups. We let their external direct product be the set

$$G = G_1 \times G_2 \times \cdots \times G_n = \{(g_1, g_2, \ldots, g_n) \mid g_i \in G_i \forall i \text{ s.t. } 1 \le i \le n\} \quad (2.2)$$

equipped with co-ordinatewise multiplication. The identity element of G is $(1, 1, \ldots, 1)$ and the inverse of $(g_1, g_2, \ldots, g_n)$ is $(g_1^{-1}, g_2^{-1}, \ldots, g_n^{-1})$. There is an internal version too of course.

Definition 2.37

Let H be a group containing normal subgroups $H_1, H_2, \ldots, H_n$. We say that H is the (internal) direct product of these subgroups H_i if the following conditions are satisfied.

(i) $H_1 H_2 \cdots H_n = H$.

(ii) For each i in the range $1 \le i \le n$, let $\widehat{H}_i$ be the product of all the H_j other than H_i, so $\widehat{H}_i = H_1 \cdots H_{i-1} H_{i+1} \cdots H_n$. We require that $H_i \cap \widehat{H}_i = 1$ for every i.

Proposition 2.38

In the notation of Definition 2.37, let E be the external direct product

$$E = H_1 \times H_2 \times \cdots \times H_n.$$

The map $\zeta : E \to H$ defined by $(h_1, h_2, \ldots, h_n) \mapsto h_1 h_2 \ldots h_n$ is an isomorphism of groups.

Proof

Using the trick with commutators outlined after Definition 2.36, we see that if i, j are different, then elements drawn from H_i commute with elements drawn from H_j.

Now define a map $\theta : E \to H$ by $(h_1, h_2, \ldots, h_n) \mapsto h_1 h_2 \cdots h_n$. The fact that each h_i, h_j from different factors commute in H is enough to ensure that

θ is a homomorphism. Now θ is surjective by part (i) of Definition 2.37. If we are working with finite groups then we can deduce that θ is bijective because domain and codomain have the same size (why?), but in general we have a little more work to do.

Suppose that $(k_1, k_2, \ldots, k_n) \in \mathrm{Ker}\,\theta$. Thus $1 = k_1 k_2 \cdots k_n \in H$. Now $k_1 = (k_2 k_3 \cdots k_n)^{-1} \in H_1 \cap \widehat{H}_1 = \{1\}$ so $k_1 = 1$. Now

$$1 = (k_1 k_2 \cdots k_n)^{k_1 k_2 \cdots k_{i-1}} = k_i k_{i+1} \cdots k_n k_1 \cdots k_{i-1}$$

for every i in the range $2 \leq i \leq n$. The argument which showed that $k_1 = 1$ will now show that $k_i = 1$ for all i in the range $2 \leq i \leq n$. Therefore $\mathrm{Ker}\,\theta = 1$ is θ is an isomorphism.

$\square$

Another name for a direct product of groups is a Cartesian product of groups. You may see the the notation $\mathrm{Cart}_{i \in I} G_i$ for the the Cartesian product of a collection $\{G_i\}$ of groups.

2.7 Finite Abelian Groups

We have already investigated cyclic groups to some extent. We will now extend that investigation, because cyclic groups are the key to understanding finite abelian groups. Our first result is technical; you may find the result unsurprising, but perhaps it is amusing that a result which looks so obvious seems to require a little work!

Lemma 2.39

Let x be an element of order n in a finite cyclic group G of order nm. It follows that $G = \langle y \rangle$ for $y \in G$ with the property that $y^m = x$.

Proof

Suppose that $G = \langle g \rangle$, so the order of g^i is $mn/\mathrm{g.c.d.}(i, mn)$ by Proposition 1.46. If $x = g^k$, then $\mathrm{g.c.d.}(k, mn) = m$. Thus k is divisible by m and so $x = g^{mj} = (g^j)^m$ where $\mathrm{g.c.d.}(mj, mn) = m$ $\mathrm{g.c.d.}(j, n) = m$, so j and n must be coprime. However, it may be that j and m are not coprime, in which case g^j will not have order mn and in consequence will not generate G.

We resolve this difficulty by induction on the number of prime factors of m (including multiplicity). If $m = 1$ there is nothing to prove. Suppose that $m = p$

is prime. If j and p are coprime we are done, so we may suppose that j and p are not coprime, so p divides j. However, j and n are coprime, so p does not divide n and so does not divide $j+n$. Thus $j+n$ is coprime to p and n and thus coprime to $pn = mn$. Now g^{j+n} is a generator of G and $(g^{j+n})^m = g^{jm}g^{mn} = x$ as required.

Finally we address the case that $m > 1$ is not prime. Choose a prime q dividing m, so G has subgroup H of order qn which contains $\langle x \rangle$ by part (ii) of Proposition 1.45. We have just shown that there is z a generator of H such that $z^q = x$. Now m/q has fewer prime factors (including multiplicity) than m, so by induction there is y a generator of G such that $y^{m/q} = z$ so $y^m = z^q = x$. The proof is complete.

$\square$

There is a faster proof of this result if you are allowed to use Dirichlet's theorem on primes numbers occurring in arithmetic progressions. Find out about this result, and see if you can use it to shorten the proof. However, Dirichlet's theorem is quite deep, and it is poor style to deploy an excessively powerful result to obtain an elementary one (the *Vorschlaghammer-Nuß* error).

Proposition 2.40

Suppose that A is a finite abelian group, and that $\max\{o(x) \mid x \in A\} = n$. It follows that if $a \in A$, then $a^n = 1$.

Proof

Choose $b \in A$ of order n. Select an arbitrary $a \in A$ so the order of a is $m \leq n$, and consider the group $H = \langle a, b \rangle \leq A$. Throughout the discussion b is fixed, though we may have occasion to tinker with a so we reserve the right to replace a by another element of order m. Euclid's algorithm yields that there are $r, s \in \mathbb{Z}$ such that $rn + sm = \text{g.c.d.}(n, m)$. We focus on the element $a^r b^s$. Let this element have order λ. We have $(a^r b^s)^\lambda = 1$ so $a^{r\lambda}$ and $b^{s\lambda}$ are both in $C = \langle a \rangle \cap \langle b \rangle$ and are mutually inverse. Let $c = |C|$ so $c \mid m, n$ by Lagrange's theorem.

The element $b^{n/c}$ has order c and so is in C by part (ii) of Proposition 1.45. We are now ready to deploy Lemma 2.39 in the cyclic group $\langle a \rangle$. This is a key point in this argument so we make a fuss. We may suppose that $a^{m/c} = b^{n/c}$, since if this is not already the case, we replace a by another generator of $\langle a \rangle$ which does enjoy this property.

Now $(a^r b^s)^\lambda = 1$ if and only if $a^{r\lambda} b^{s\lambda} = 1$. Now $a^{r\lambda}, b^{s\lambda} \in C$ so $r\lambda = u(m/c)$, $s\lambda = v(n/c)$ for $u, v \in \mathbb{Z}$. Thus

$$(a^r b^s)^\lambda = a^{u(m/c)} b^{v(n/c)} = (a^{m/c})^u (b^{n/c})^v = (a^{m/c})^{u+v}$$

since we have arranged that $a^{m/c} = b^{n/c}$. The condition that $(a^r b^s)^\lambda = 1$ simply becomes the condition that c divides $u + v$. Now $rn\lambda = u(mn/c)$ and $sm\lambda = v(mn/c)$ and therefore $(rn+sm)\lambda = mn(u+v)/c$ with $(u+v)/c \in \mathbb{Z}$, and since each side is positive, we even have $(u+v)/c \in \mathbb{N}$. Thus g.c.d.$(n, m)\lambda \geq mn$ so $\lambda \geq (m/\text{g.c.d.}(n, m))n \geq n$. Now λ is the order of an element of A, and the maximality of n ensures that $\lambda = n$, so $m = $ g.c.d.(n, m) and therefore m divides n.

$\square$

Lemma 2.41

Suppose that A is a finite abelian group and that $b \in A$ is of maximal order n. It follows that either A is cyclic, or A has a subgroup $H \neq 1$ such that $H \cap \langle b \rangle = 1$.

Proof

If A is not cyclic, we choose $x \in A - \langle b \rangle$ and suppose that x has order m. Suppose that $\langle x \rangle \cap \langle b \rangle = \langle b^{n/c} \rangle$ is a group of order c. As in the proof of Proposition 2.40, we may replace x by a slight improvement. Thanks to Lemma 2.39 we have $\langle x \rangle = \langle a \rangle$ where $a^{m/c} = b^{n/c}$. Now by Proposition 2.40, m divides n so m/c divides n/c. Note that $b^{n/c} = (b^{n/m})^{m/c}$. Now $\langle x, b \rangle = \langle a, b \rangle = \langle a(b^{n/m})^{-1}, b \rangle$. Let $h = a(b^{n/m})^{-1}$ so $h^{m/c} = 1$. Moreover $a^i \notin \langle b \rangle$ for $1 \leq i < m/c$ so $o(h) = m/c$ and $\langle h \rangle \cap \langle b \rangle = 1$.

$\square$

We now introduce some useful terminology.

Definition 2.42

Suppose that G is a group and that A is a normal subgroup of G. A *complement* for A in G is a subgroup B of G such that $G = AB$ and $A \cap B = 1$.

In our abelian context, normality is not an issue, and a complement will give rise to an internal direct product.

Theorem 2.43

Every finite abelian group A is a direct product of cyclic groups. Any cyclic subgroup of maximal order can occur as one of the factors in such a decomposition.

Proof

Choose $b \in A$ of maximal order. By Lemma 2.41 there is $1 \neq H \leq A$ such that $H \cap \langle b \rangle = 1$. Note that Hb has maximal order in A/H, so by induction on group order we have $A/H = L \times \langle Hb \rangle$ for some $L \leq A/H$. Let $L = M/H$ for $H \leq M \leq A$ so $M \cap \langle b \rangle \leq H$ and therefore $M \cap \langle b \rangle \leq H \cap \langle b \rangle = 1$. Moreover, it is clear that $M \langle b \rangle = A$. Thus M is a complement for $\langle b \rangle$.

By induction on group order M is a finite direct product of cyclic groups and we are done.

$\square$

In fact we can squeeze out a little more.

Theorem 2.44

Suppose that G is a finite abelian group, then

(a)
$$G \simeq C_{d_1} \times C_{d_2} \times \cdots \times C_{d_t} \tag{2.3}$$

where $1 < d_i \mid d_{i+1}$ for every i in the range $1 \leq i < t$.

(b) Moreover, this number t and the sequence of numbers (d_i) are uniquely determined by G.

Proof

The proof of part(a) is by induction on $|G|$. The case $|G| = 1$ corresponds to $t = 0$ (no factors).

Suppose that $|G| > 1$. The fact that such a decomposition exists follows from previous work, since we have shown that if $x \in G$ has maximal order, then there is $L \leq G$ such that $G = L \times \langle x \rangle$ (an internal direct product), and part (a) of the result is true for L by induction. Now apply Proposition 2.40 to see that $d_{t-1} \mid d_t = o(x)$ and so we have a decomposition of G of the form proposed in (2.3).

We now approach part (b). Choose $d \in \mathbb{N}$ and let $G_d = \{g \mid g \in G, g^d = 1)$. Note that $\log_d |G_d|$ is maximal if and only if $d \mid d_1$ and for such a d, $\log_d |G_d| = t$.

Now d_1 is determined by G, since d_1 is the largest value of d for which $\log_d |G_d|$ is maximal. In fact we can also recover t as this stage via $t = \log_{d_1} |G_{d_1}|$.

In our decomposition $G = A_1 \times A_2 \times \cdots \times A_t$ where $A_i \simeq C_{d_i}$ for every i, each factor A_i has a unique subgroup $B_i \simeq C_{d_1}$. There are natural projections $\theta_i : A_i \to L_i = A_i/B_i \simeq C_{d_i/d_1}$, and these induce a homomorphism $\theta : G \to L = L_1 \times L_2 \times \cdots \times L_t$ with kernel G_{d_1}. Thus $\mathrm{Im}\,\theta \simeq G/G_{d_1}$ but $|G/G_{d_1}| = |G : G_{d_1}| = |G|/d_1^t = |L|$ so θ is surjective. Moreover, we have a decomposition of L as a direct product of cyclic groups L_i with $|L_i|$ dividing $|L_{i+1}|$ for all i. Discarding the trivial factor(s) at the beginning, we have a factorization for L into cyclic group as guaranteed by part (a) of this theorem. By induction on group order, the order $|L_i|$ of each factor L_i (including the trivial ones) is uniquely determined by G/G_{d_1} (and therefore by G).

We already knew that G determined d_1 and t, but now we also know that it determines d_i/d_1 for each i, and therefore determines each d_i. The proof is complete.

$\square$

Order	Isomorphism types of abelian group
1	The trivial group
2	C_2
3	C_3
4	C_4, $C_2 \times C_2$
5	C_5
6	C_6
7	C_7
8	C_8, $C_2 \times C_4$, $C_2 \times C_2 \times C_2$
9	C_9, $C_3 \times C_3$
10	C_{10}
11	C_{11}
12	C_{12}, $C_2 \times C_6$
13	C_{13}
14	C_{14}
15	C_{15}
16	C_{16}, $C_2 \times C_8$, $C_4 \times C_4$, $C_2 \times C_2 \times C_4$, $C_2 \times C_2 \times C_2 \times C_2$
17	C_{17}
18	C_{18}, $C_3 \times C_6$
19	C_{19}
20	C_{20}, $C_2 \times C_{10}$

This wonderful theorem enables us to classify all finite abelian groups up to isomorphism. We give the smallest ones above.

EXERCISES

2.14 Are $C_{19} \times C_{91}$ and $C_{13} \times C_{133}$ isomorphic groups?

2.15 Let p be a prime number. Show that the number of isomorphism-types of finite abelian groups of order p^n which can be generated a two element subset is $\lfloor n/2 + 1 \rfloor$ (here $\lfloor x \rfloor$ is the integer-part of $x \in \mathbb{R}$).

2.16 Let p be a prime number. Show that the number of isomorphism-types of finite abelian groups of order p^n which can be generated a three element subset is $\lfloor \frac{n^2+6n+12}{12} \rfloor$ (using the same notation as in Question 1).

2.17 Show that one cannot omit the condition that G is abelian in Proposition 2.40.

2.8 Finitely Generated Abelian Groups

In the previous section we nailed down finite abelian groups very firmly. We now tackle the more ambitious project of trying to understand those abelian groups which may be generated by a finite subset of their elements. In fact we shall succeed once again. The moral is, if you want an abelian group to be mysterious, then make sure that it is not finitely generated.

Proposition 2.45

Suppose that $G = \langle x_1, \ldots, x_n \rangle$ is a finitely generated abelian group and that H is a subgroup of G. It follows that H can be generated by n elements.

Proof

We proceed by induction on $n \geq 0$. When $n = 0$ both G and H are trivial groups, and there is nothing to do. We may assume that $n \geq 1$. Let $N = \langle x_1, \ldots x_{n-1} \rangle$. Now G is abelian so $N \trianglelefteq G$. The group $H \cap N$ can be generated by $n-1$ elements by inductive hypothesis, and $H/H \cap N \simeq HN/N \leq G/N$ by the second isomorphism theorem.

Since G is abelian, every element of G can be written in the form $x_1^{\alpha_1} \cdots x_n^{\alpha_n}$ where the exponents are integers. Each $x_1^{\alpha_1} \cdots x_{n-1}^{\alpha_{n-1}} \in N$ so every coset of N in G is of the form $N x_n^{\alpha_n}$. Moreover, as α_n ranges over the integers each $N x_n^{\alpha_n}$ is a right coset of N in G. Thus $G/N = \langle N x_n \rangle$ is a cyclic group. Now HN/N is a subgroup of a cyclic group, and so is cyclic. Thus $H/H \cap N = \langle (H \cap N)y \rangle$ is a cyclic group. It follows that every element H is an element of $H \cap N$ multiplied on the right by a power of y. We adjoin y to a generating set of size at most $n - 1$ for $H \cap N$ to obtain a generating set of size at most n for H.

$\square$

This result is quite strong. It is true that a subgroup of a finite group must be finite, but a subgroup of a finitely generated non-abelian group G need not be finitely generated, let alone be generated by a set of the same size as a generating set for G. We shall construct an explicit example of this phenomenon. See Remark 4.18.

In an abelian group G, it is straightforward to check that the elements of finite order form a subgroup. This subgroup is called the torsion subgroup $\tau(G)$. If G is also finitely generated, then Proposition 2.45 tells us that $\tau(G)$ is a finitely generated abelian group consisting of elements of finite order. This forces $\tau(G)$ to be a finite abelian group, and all the results of the previous section apply. If G has the property that $\tau(G) = 1$, we say that G is *torsion-free*. Another way of saying that G is torsion-free is to say that the identity is the only element of G which has finite order.

Proposition 2.46

Suppose that G is a finitely generated abelian group, then $G/\tau(G)$ is a torsion-free finitely generated abelian group.

Proof

Suppose that $x_1, \ldots, x_n$ generate G, and that $\pi : G \to G/\tau(G)$ is the natural epimorphism. The fact that π is surjective ensures that $G/\tau(G)$ is generated by $\{(x_1)\pi, \ldots, (x_n)\pi\}$, and so is finitely generated. Now suppose that $\tau(G)y$ is an element of finite order m in $G/\tau(G)$. Now $\tau(G) = (\tau(G)y)^m = \tau(G)y^m$ so y^m has finite order t. Thus $y^{mt} = (y^m)^t = 1$ so $y \in \tau(G)$ and $\tau(G)y = \tau(G)$. The only element of $G/\tau(G)$ which has finite order is the identity, so $G/\tau(G)$ is torsion-free.

$\square$

The *generating rank* of an abelian group G is the minimum size of a generating set for G. The trivial group has generating rank 0, non-trivial cyclic groups have generating rank 1. Note that $C_{42} \times C_\infty$ can be generated by two elements. It is not the trivial group so its generating rank is not 0. It is not a finite cyclic group since it contains an element of infinite order, and it is not an infinite cyclic group since it contains an element of order 42. Thus its generating rank is not 1. Thus its generating rank is 2, as the reader may have guessed. However, $C_2 \times C_3$ has generating rank 1 because $C_2 \times C_3 \simeq C_6$. Note that the rational numbers under addition are not finitely generated, and so that group has infinite generating rank.

Theorem 2.47

Suppose that G is torsion-free finitely generated abelian group of generating rank n.

(a)

$$G \simeq \underbrace{C_\infty \times C_\infty \times \cdots \times C_\infty}_{n \text{ copies of } C_\infty}.$$

(b) If H is a subgroup of G and $|G : H| < \infty$, then $H \simeq G$.

Proof

We prove these result by induction on $n \geq 0$. First note that the results are true when $n = 0$, and the more timid may even note that they are true when $n = 1$, but this is an unnecessary observation. We assume that $n \geq 1$. Choose a generating set $\{x_1, \ldots, x_n\}$ of size n for G.

Suppose (for contradiction) that there is a non-trivial relation of the form $x_1^{\alpha_1} \cdots x_n^{\alpha_n} = 1$. Relabelling if necessary, we may assume that $\alpha_n \neq 0$. Now

$$x_n^{\alpha_n} \in L = \langle x_1, \ldots, x_{n-1} \rangle,$$

so $1 < t = |G : L| \leq |\alpha_n| < \infty$. Notice that L must have minimum generating rank $n - 1$, else we could find a generating set of size less than n for G. Now by induction, L is the direct product of $n - 1$ copies of C_∞.

Consider the map $\psi : G \to L$ defined by $(g)\psi = g^t$. Certainly each $(g)\psi \in L$ by Lagrange's theorem applied to G/L. Also for every $a, b \in G$ we have

$$(ab)\psi = (ab)^t = a^t b^t = (a)\psi(b)\psi$$

so ψ is a homomorphism. Moreover, $\text{Ker}(\psi) = 1$ since G is torsion free, so $M = \text{Im } \psi \simeq G$ by the first isomorphism theorem. Thus $G \simeq \text{Im } \psi \leq L \leq G$. Now $G/\text{Im } \psi$ is finitely generated since G is finitely generated. Moreover it is abelian since G is abelian, and has the property that $x^t = 1$ for every $x \in G/\text{Im } \psi$. Thus $|G : \text{Im } \psi|$ is finite so $|L : \text{Im } \psi| < \infty$ by Corollary 1.30.

Now by induction $G \simeq \operatorname{Im} \psi \simeq L$ and so G which is of generating rank n can be generated by $n-1$ elements. This is absurd and so there is no non-trivial relation among the x_i where $1 \le i \le n$. It follows that G is the internal direct product of the infinite cyclic groups $\langle x_i \rangle$ and so $G \simeq C_\infty \times \cdots \times C_\infty$. We have established part (a).

We can finish very quickly. Suppose that H has finite index in G. By Proposition 2.45 it follows that H has at most n generators. If (for contradiction) H had generating rank less than n, then we would obtain a contradiction by the argument in the previous paragraph. Thus the generating rank of H is n, and so part (b) follows.

$\square$

Note that if the torsion-free abelian group G has generating rank n, then

$$G \simeq C_\infty^n = \underbrace{C_\infty \times C_\infty \times \cdots \times C_\infty}_{n \text{ copies of } C_\infty}$$

It is also true, but requires proof, that for every natural number m the group C_∞^m has generating rank m. In case this seems strange, note that we have yet to demonstrate that if $C_\infty^n \simeq C_\infty^m$, then $n = m$. However, this is not hard. Suppose that $C = C_\infty^n$, and let $C^{[2]} = \{c^2 \mid c \in C\} \le C$. There is a unique epimorphism $\eta : C_\infty \to C_2$, and the kernel of this map consists of the squares in C. Define a map $\varphi : C \to C_2^n$ via $(x_1, \ldots, x_n) \mapsto ((x_1)\eta, \ldots, (x_n)\eta)$. Now φ is visibly an epimorphism with kernel $C^{[2]}$. Thus $|C : C^{[2]}| = |\operatorname{Im} \varphi| = 2^n$ by the first isomorphism theorem. Thus n is determined by $|C : C^{[2]}|$ and therefore by C. It follows that the generating rank of C_∞^n is n.

Definition 2.48

Suppose that A is an abelian group and $X \subseteq A$. We say that A is a *free abelian group on* X if whenever B is an abelian group, and $\theta : X \to B$ is a map, there exists a unique group homomorphism $\overline{\theta} : A \to B$ such that $(x)\theta = (x)\overline{\theta}$ for every $x \in X$.

Note that if the abelian group G is the internal direct product of finitely many infinite cyclic subgroups $\langle x_i \rangle$ where $1 \le i \le n$, we can put $X = \{x_i \mid 1 \le i \le n\}$ and it follows that G is free on X. This is because each element of G is uniquely expressible as $x_1^{\alpha_1} \cdots x_n^{\alpha_n}$ so any homomorphism $G \to B$ which agrees with θ on X must send each $x_1^{\alpha_1} \cdots x_n^{\alpha_n}$ to $((x_1)\theta)^{\alpha_1} \cdots ((x_n)\theta)^{\alpha_n}$. However, this rule defines a perfectly satisfactory homomorphism from G to B whenever all values of $(x_i)\theta$ are prescribed, and this is the map $\overline{\theta}$. Now G is free abelian on the set X.

Proposition 2.49

Suppose that A is a free abelian group on X, and that $\nu : G \to A$ is an epimorphism of abelian groups. It follows that G has a subgroup H which is isomorphic to A, and is such that $G = \mathrm{Ker}\, \nu \times H$.

Proof

For each $x \in X$ choose $g_x \in G$ such that $(g_x)\nu = x$. Let $H = \langle g_x \mid x \in X \rangle$. Define a map $\zeta : X \to H$ by $\zeta : x \mapsto g_x$ for every $x \in X$. Now use the free abelian property; there is a unique homomorphism $\bar{\zeta} : A \to H$ which agrees with ζ on X. The definitions ensure that $\bar{\zeta} : A \to H$ is surjective. It remains to verify that H has the required properties. Let $\bar{\nu}$ denote the restriction of ν to H. Thus $\bar{\nu} : H \to A$ and $(h)\bar{\nu} = (h)\nu \; \forall h \in H$.

Notice that $(x)(\bar{\zeta} \circ \bar{\nu}) = (g_x)\nu = x$ for each $x \in X$. Another homomorphism from A to A which sends each $x \in X$ to itself is the identity map. By the uniqueness clause in the definition of being free abelian on X, it follows that $\bar{\zeta} \circ \bar{\nu} = \mathrm{Id}_A$. The map $\bar{\nu}$ is a right inverse of $\bar{\zeta}$ and so $\bar{\zeta}$ is injective. We already know that $\bar{\zeta}$ is surjective, so $\bar{\zeta}$ is an isomorphism and $H \simeq A$ as required. Moreover $\bar{\nu}$ is the two-sided inverse of $\bar{\zeta}$.

Also note that if $y \in \mathrm{Ker}\, \nu \cap H$, then $(y)\bar{\nu} = 1$ and therefore

$$ y = (y)(\bar{\nu} \circ \bar{\zeta}) = ((y)\bar{\nu})\bar{\zeta} = (1)\bar{\zeta} = 1. $$

Thus $\mathrm{Ker}\, \nu \cap H = 1$.

To complete the argument we must show that $(\mathrm{Ker}\, \nu)H = G$. Choose any $g \in G$, then $g = z \cdot ((g)\nu)\bar{\zeta}$ where z is some element of G. Now

$$ (g)\nu = (z)\nu \cdot (((g)\nu)\bar{\zeta})\nu = (z)\nu \cdot (g)\nu $$

so $z \in \mathrm{Ker}\, \nu$ and of course $((g)\nu)\bar{\zeta} \in H$ as required.

$\square$

Theorem 2.50

Suppose that G is a finitely generated abelian group, then G can be expressed as a direct product

$$ C_{d_1} \times \ldots C_{d_k} \times C_\infty^m $$

where $1 < d_1 \mid d_2 \mid \cdots \mid d_k$, and the numbers $k, m, d_1, \ldots, d_k$ are uniquely determined by G.

Proof

We have assembled the tools to do this quickly. The torsion subgroup $\tau(G)$ is finite, and the structure theorem for finite abelian groups applies. The factor group $G/\tau(G)$ is a finitely generated and torsion-free abelian group. It is therefore a free abelian group by Theorem 2.47, and so by Proposition 2.49 there is $H \leq G$ such that $G = \tau(G) \times H$ and H is a finitely generated free abelian group. The generating rank m of H is the generating rank of $G/\tau(G)$, and this is determined by G.

$\square$

2.9 Semi-direct Products

A group H is the internal direct product of its subgroups A and B if the following conditions all hold: $A, B \trianglelefteq G$, $AB = G$ and $A \cap B = 1$. We tinker with this definition just a little.

Definition 2.51

We say that G is the semidirect product of A by B if the following conditions all hold: $A \trianglelefteq G$, $B \leq G$, $AB = G$ and $A \cap B = 1$. Thus we have relaxed the condition that B be normal. We write $G = A \rtimes B$.

Suppose that G is the semidirect product of A by B. It is still the case that every $g \in G$ has a unique expression as $g = ab$ with $a \in A$ and $b \in B$. This is because if $g = a_1 b_1$ were a rival expression with $a_1 \in A, b_1 \in B$ then $a_1^{-1} a = b_1 b^{-1} \in A \cap B = \{1\}$. Thus $a = a_1$ and $b = b_1$. In an internal direct product, elements coming from different factors must commute. This is not the case in a semidirect product. For example, in S_3 the symmetric group on $\{1, 2, 3\}$, there is a normal subgroup $C = \langle (1, 2, 3) \rangle$ and another subgroup $D = \langle (1, 2) \rangle$. It is straightforward to check that S_3 is the semidirect product of C by D but $(1, 2)$ and $(1, 2, 3)$ do not commute.

The existence of $N \trianglelefteq G$ does not guarantee the existence of $H \leq G$ such that G is the semidirect product of N by H. An example of this arises in the cyclic group of order 4. Consider the multiplicative group of complex numbers $M = \{1, -1, i, -i\}$. Now M has a normal subgroup $P = \{1, -1\}$, but there is no subgroup Q such that M is the semidirect product of P by Q. To see this, note that M only has three subgroups: the trivial group, P and M itself, and none of them satisfies the necessary conditions. However, given a normal subgroup N of G, if there is $H \leq G$ such that G is the semidirect product of N by H, then we say that H is a complement for N in G. The first isomorphism theorem tells us that $H \simeq H/H \cap N \simeq HN/N = G/N$. We also say that G *splits* over N.

It is worth looking at how multiplication works in a semidirect product of N by H; suppose that $h_1, h_2 \in H$ and $n_1, n_2 \in N$ then

$$(h_1 n_1)(h_2 n_2) = h_1 h_2 h_2^{-1} n_1 h_2 n_2 = (h_1 h_2)(n_1^{h_2} n_2).$$

Thus in order to know how to multiply elements in $N \rtimes H$, we need to know the multiplication in H and N, and how the elements of H conjugate elements of N. Note that if elements of H commute with elements of N, then we actually have an ordinary direct product on our hands.

Now suppose that instead of having a group G which happens to be a semidirect product of N by H, we are instead given the groups H and N and seek to construct G. In fact you need an extra ingredient, because you need to know how elements of H conjugate elements of N. Let Aut N denote the automorphism group of N, this being the group consisting of the isomorphisms from N to N, with composition of maps being the group operation. The extra ingredient we need to capture the conjugation is a homomorphism $\varphi : H \to$ Aut N.

Now as a set, we let W be $H \times N$. However, the multiplication we impose on W is not necessarily co-ordinatewise as in a direct product. The recipe is this:

$$(h_1, n_1) \star (h_2, n_2) = (h_1 h_2, (n_1)(h_2\varphi)n_2) \ \forall h_1, h_2 \in H, \ \forall n_1, n_2 \in N$$

or using the suggestive exponential notation for automorphisms we have

$$(h_1, n_1) \star (h_2, n_2) = (h_1 h_2, (n_1)^{(h_2\varphi)} n_2) \ \forall h_1, h_2 \in H, \ \forall n_1, n_2 \in N$$

If φ happens to be the uninteresting homomorphism which sends each h to the identity automorphism, then we obtain a perfectly ordinary direct product. However, if φ is not trivial, we have incorporated a twist in the multiplication which makes life a deal more colourful.

The operation on W is certainly closed, and $(1, 1)$ is a two-sided identity. The inverse of (h, n) is $(h^{-1}, (n^{-1})((h^{-1})\varphi))$. We must perform a penance and check the associative law. Suppose that $(h_i, n_i) \in W$ for $i = 1, 2, 3$. Thus

$$
\begin{aligned}
((h_1, n_1) \star (h_2, n_2)) \star (h_3, n_3) &= (h_1 h_2, n_1^{(h_2)\varphi} n_2 \star (h_3, n_3) \\
&= (h_1 h_2 h_3, (n_1^{(h_2)\varphi} n_2)^{(h_3)\varphi} n_3) \\
&= (h_1 h_2 h_3, n_1^{(h_2 h_3)\varphi} n_2^{(h_3)\varphi} n_3).
\end{aligned}
$$

On the other hand

$$
\begin{aligned}
(h_1, n_1) \star ((h_2, n_2) \star (h_3, n_3)) &= (h_1, n_1) \star (h_2 h_3, n_2^{(h_3)\varphi} n_3) \\
&= (h_1 h_2 h_3, n_1^{(h_2 h_3)\varphi} n_2^{(h_3)\varphi} n_3)
\end{aligned}
$$

so the associative law holds.

Of course H and N are not subgroups of W, but they embed monomorphically into W via $h \mapsto (h, 1)$ and $n \mapsto (1, n)$. We call the image groups $\overline{H}$ and $\overline{N}$ respectively . Now you can verify that W is the internal semidirect product of $\overline{N}$ by $\overline{H}$.

We will not use wreath products in the sequel. The final part of this chapter may be omitted during a first reading.

2.10 Wreath Products

Suppose that A and B are groups, and that A acts on a set Ω. Form the group $C = \mathrm{Cart}_{\omega \in \Omega} B$. There are various ways to think about C. There is the option of thinking of C as consisting of "sequences" (b_ω) which are supposed to be $|\Omega|$-tuples of elements of B. We are stretching the meaning of sequences here since the Ω may not be countable. Elements of C are composed co-ordinatewise. It is more honest to think of C as consisting of all possible functions from Ω to B, with the group law $*$ being $f_1 * f_2 : \omega \mapsto (\omega)f_1 \cdot (\omega)f_2$ whenever $f_1, f_2 : \Omega \to B$.

There is a subgroup C_f of C consisting of "sequences" or more honestly functions of finite support. In sequence language, this means all except for finitely many terms are 1, and in function terms this means that we focus on those functions f which have the property that $(\omega)f = 1$ for all except for finitely many values of ω. Of course, if Ω is finite then $C = C_f$.

We have an action of A on C defined by $(b_\omega) \cdot a = (\overline{b}_\omega)$ where for each $\omega \in \Omega$ we have

$$
\overline{b}_\omega = b_{\omega \cdot a^{-1}} \tag{2.4}
$$

where the dot in the subscript denotes the action of A on Ω. All that happens is that the entries of (b_ω) are being permuted according to the action of a on their positions. The appearance of a^{-1} rather than a in Eq. (2.4) should be greeted with fervent equanimity. The reason it is there is because you are moving the terms of the sequence (b_ω) around according to the action of a, but we are describing this motion via the labels of the positions, not via the terms. Thus after you move the contents according to the action of a, the entry in position ω must have previously resided in position $\omega \cdot a^{-1}$. When you come to terms with this, Eq. (2.4) loses all mystery.

We have an induced homomorphism $\varphi : A \to \mathrm{Aut}\, C$, and can therefore form the semidirect product W of C with A. This is called the (upper case W) *Wreath* product of B with A with base group B. If you play a very similar game, but use C_f instead of C, you get the (lower case w) wreath product W_f of B with A. In the event that Ω is finite, the two notions coincide. If Ω is infinite, the lower case wreath product will be a proper subgroup of the upper case Wreath product.

Every element of the wreath or Wreath product will be an ordered pair (a, c) with $a \in A$ and c in the appropriate group, and the product law will be

$$\left(a_1, (b_\omega^1)\right) \cdot \left(a_2, (b_\omega^2)\right) = \left(a_1 a_2, (b_{\omega \cdot a_1^{-1}}^1)(b_\omega^2)\right).$$

A concrete example is in order.

Example 2.52

We form the wreath product of the infinite cyclic group $\langle g \rangle$ (as base group) and the subgroup $\langle \gamma \rangle$ of S_3 where $\gamma = (1, 2, 3)$. Thus $\langle \gamma \rangle$ acts naturally on $\Omega = \{1, 2, 3\}$. Now C consists of triples (b_1, b_2, b_3) where $b_1 = g^l$, $b_2 = g^m$ and $b_3 = g^n$. Let us calculate a sample product in the wreath product. We have

$$
\begin{aligned}
\left(\gamma^2, (g^2, g^{-3}, g^0)\right) \cdot \left(\gamma, (g^5, g^7, g^{11})\right) &= \left(\gamma^2\gamma, (g^2, g^{-3}, g^0)^\gamma (g^5, g^7, g^{11})\right) \\
&= \left(\gamma^3, (g^0, g^2, g^{-3}) (g^5, g^7, g^{11})\right) \\
&= \left(1, (g^5, g^9, g^8)\right).
\end{aligned}
$$

Let us return to the general case. We will work with the Wreath product and make parenthetical remarks about the wreath product). Now we know how multiplication works, it is clear that $\theta : W \to A$ defined by $(a, c) \mapsto a$ is a group epimorphism with kernel C (or C_f). Thus C (or C_f) is a normal subgroup of W (or W_f). There is an embedding (i.e. a monomorphism) $A \to W$ or W_f defined by $a \mapsto (a, 1)$.

3

Action

3.1 Permutation Groups

Let Ω be a set. Recall that the symmetric group Sym Ω on Ω consists of all bijections from Ω to Ω, with map composition as the group operation. An important special case is when Ω is finite and consists of the first n natural numbers. We refer to this symmetric group as S_n. If Γ is any set of size n, then there is a bijection $\theta : \Gamma \to \{1, 2, \ldots, n\}$ which induces a map $\bar{\theta} :$ Sym $\Gamma \to S_n$ via $(\sigma)\bar{\theta} = \theta^{-1} \circ \sigma \circ \theta$ for every $\sigma \in$ Sym Γ. It is straightforward to verify that $\bar{\theta}$ is a bijective homomorphism, and so an isomorphism of groups. This fits our intuition very nicely; if we are studying the bijections from Ω to Ω it does not matter what the elements of Ω actually are! Now we know this, there is a powerful temptation to describe Sym Γ as S_n, just as C_n gets used to mean the cyclic group of order n that we happen to be thinking about at a given moment.

Definition 3.1

A *permutation group* G on a set Ω is a subgroup of Sym Ω. The *degree* of this permutation group is $|\Omega|$.

Suppose that $\sigma \in$ Sym Ω. We define the *support* of σ to be the collection of elements of Ω which are not fixed points of σ. Thus we write

$$\operatorname{supp}(\sigma) = \{\omega \mid \omega \in \Omega, \ (\omega)\sigma \neq \omega\}.$$

If Ω is an infinite set, the support of a particular σ may be infinite. If the support of σ is finite, we say that σ is a finitary permutation (because it is rather like a permutation of a finite set). Of course the identity permutation has finite support because its support is the empty set. The product (composite) of two finitary permutations is finitary. Moreover $\operatorname{supp}(\sigma) = \operatorname{supp}(\sigma^{-1})$ for any $\sigma \in \operatorname{Sym} \Omega$, so the inverse of a finitary permutation is finitary. Thus the set of finitary permutations forms a subgroup of $\operatorname{Sym} \Omega$. Denote this group of finitary permutations by $\operatorname{Sym}_f \Omega$. Thus

$$\operatorname{Sym}_f \Omega = \{\sigma \mid \sigma \in \operatorname{Sym} \Omega, |\operatorname{supp}(\sigma)| < \infty\}$$

is a permutation group on Ω. Note that $\operatorname{Sym}_f \Omega = \operatorname{Sym} \Omega$ if and only if Ω is finite.

EXERCISES

3.1 Suppose that Ω is a set. Which of the following are subgroups of $\operatorname{Sym} \Omega$?

(a) $\{\sigma \mid \sigma \in \operatorname{Sym} \Omega, |\operatorname{supp}(\sigma)| \text{ is infinite }\}$

(b) $\{\sigma \mid \sigma \in \operatorname{Sym} \Omega, |\operatorname{supp}(\sigma)| \text{ is countable }\}$

(c) $\{\sigma \mid \sigma \in \operatorname{Sym} \Omega, |\operatorname{supp}(\sigma)| < 2\}$

(d) $\{\sigma \mid \sigma \in \operatorname{Sym} \Omega, |\operatorname{supp}(\sigma)| < 3\}$.

3.2 Show that the symmetric group S_{2n} has two subgroups A, B with disjoint support, each of A and B being isomorphic to S_n. deduce that S_{2n} has a subgroup isomorphic to $S_n \times S_n$.

The elements of a permutation group move the elements of Ω. The primary concern of a footloose element of Ω must surely be: *where can I go?*. To address this important question, we introduce the notion of an orbit.

Definition 3.2

Suppose that H is a permutation group on Ω, and $\omega \in \Omega$. We define the H-orbit of ω to be

$$\omega H = \{(\omega)h \mid h \in H\} \subseteq \Omega.$$

Lemma 3.3

In the notation of Definition 3.2 suppose that $\omega' \in \Omega$. It follows that $\omega H = \omega' H$ if and only if $\omega' \in \omega H$ (and therefore if and only if $\omega \in \omega' H$).

Proof

First suppose that $wH = w'H$. Therefore $w' = w' \cdot 1 \in wH$.

We are half way done, and now address the reverse implication. Suppose that $w' \in wH$. Thus $w' = (w)h$ for some $h \in H$. It follows that

$$
\begin{aligned}
w'H &= ((w)h)H = \{((w)h)h' \mid h' \in H\} \\
&= \{(w)(hh') \mid h' \in H\} = \{(w)k \mid k \in H\} = wH.
\end{aligned}
$$

The penultimate equality is because as h' runs over H, so does hh' where h is any fixed element of H.

$\square$

For example, Ω is the Sym Ω-orbit of any $w \in \Omega$, and it is also the $\mathrm{Sym}_f \Omega$-orbit of w, since you can use an element of either of these groups to transport w to anywhere in Ω.

Definition 3.4

Suppose that Ω is a set. A *transposition* τ is an element of Sym Ω which exchanges two elements of Ω, and fixes all other elements.

Definition 3.5

The element $\sigma \in$ Sym Ω is a cycle if $\mathrm{supp}(\sigma)$ consists of a single $\langle\sigma\rangle$-orbit.

Note that a cycle may be finite or infinite. The map $\alpha : \mathbb{Z} \to \mathbb{Z}$ defined by $z \mapsto z+2$ for even z and $z \mapsto z$ for odd z is an infinite cycle in the permutation group Sym $\mathbb{Z}$. If $m \geq 2$ is a natural number, an m-cycle is a cycle with support of size m.

Note that a transposition is exactly the same thing as a 2-cycle. Here is an example of a 3-cycle ζ in S_4: we specify that $(1)\zeta = 1$, $(2)\zeta = 4$, $(3)\zeta = 2$ and $(4)\zeta = 3$. This 3-cycle is illustrated in Figure 3.1.

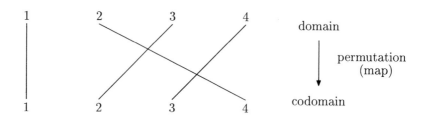

Figure 3.1 1 to 1, 2 to 4, 3 to 2 and 4 to 3

Proposition 3.6

Let $m \geq 2$ be a positive integer and suppose that $\sigma \in$ Sym Ω is an m-cycle. Thus supp $\sigma \neq \emptyset$. Choose any $\omega \in$ supp σ. It follows that the finite sequence $\omega, (\omega)\sigma, \dots, (\omega)\sigma^{m-1}$ lists the elements of $\text{supp}(\sigma)$ without repetition, and $(\omega)\sigma^m = \omega$ (so supp $\sigma = \omega\langle\sigma\rangle$).

Proof

Since $\sigma \in$ Sym Ω is an m-cycle, its support consists of a single $\langle\sigma\rangle$-orbit. Suppose that $(\omega)\sigma^i = (\omega)\sigma^j$ where $0 \leq i < j \leq m$ and $j - i$ is minimized. Thus $(\omega)\sigma^{j-i} = \omega$, from which it follows that the $\langle\sigma\rangle$-orbit of ω has size $j - i \leq m$. Now Lemma 3.3 ensures that $j - i = m$ so $i = 0$, $j = m$, and the elements $\omega, (\omega)\sigma, \dots, (\omega)\sigma^{m-1}$ are distinct.

$\square$

We can describe an m-cycle using the list mentioned in Proposition 3.6. Of course the order in which the list occurs is not unique, since we may start with any element of supp σ. The permutation ζ illustrated in Figure 3.1 can be described by the list $2, 4, 3$, the list $4, 3, 2$ and the list $3, 2, 4$. The standard notation is to place such a list in round brackets, so

$$\zeta = (2, 4, 3) = (4, 3, 2) = (3, 2, 4).$$

If a unique form of expression is essential for your purposes, the standard convention is to choose that description which minimizes the first "coordinate", so $\zeta = (2, 4, 3)$.

Composition of permutations is from left to right, and we illustrate the facts that $(1, 2)(2, 4, 3) = (1, 4, 3, 2)$ and $(2, 4, 3)(1, 2) = (1, 2, 4, 3)$ in Figure 3.2.

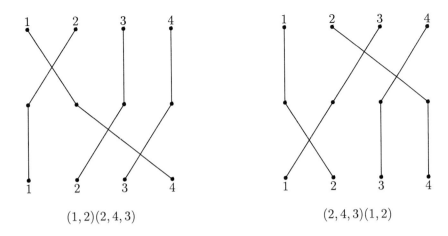

$$(1,2)(2,4,3) \qquad\qquad\qquad (2,4,3)(1,2)$$

Figure 3.2 $(1,2)$ then $(2,4,3)$ on the left, $(2,4,3)$ then $(1,2)$ on the right

An example of a 4-cycle is $(2,3,4,6)$. There is ambiguity about the group which contains ζ. It is implicit that we are working in S_n where $n \geq 6$.

Remark 3.7

It is very important to be able to compose permutations in your head. For example, consider the first product mentioned in Fig 3.2:

$$(1,2)(2,4,3) = (1,4,3,2) \qquad\qquad (3.1)$$

This calculation is performed as follows. Read the left side of Eq.(3.1) from left to right. Start with 1 and see where this lands by virtue of the application of successive cycles. The first cycle sends 1 to 2, so now you concern yourself with what the next (and in this case final) cycle does to 2. In fact 2 gets propelled to 4. One could write $1 \mapsto 2 \mapsto 4$. Thus the product begins $(1,4\ldots)$. Next we concern ourselves with the fate of 4. Now $4 \mapsto 4 \mapsto 3$ so our product becomes $(1,4,3\ldots$. The destiny of 3 is our next preoccupation. We see that $3 \mapsto 3 \mapsto 2$ so our product becomes $(1,4,3,2\ldots$ and the product must complete to $(1,4,3,2)$ since 4 must be the image of some element in $\{1,2,3,4\}$.

Example 3.8

Suppose that $1 \leq i < j \leq n$. Let $\sigma_k = (k, k+1)$ then

$$(i,j) = \sigma_{j-1}\sigma_{j-2}\cdots\sigma_{i+1}\sigma_i\sigma_{i+1}\cdots\sigma_{j-2}\sigma_{j-1}.$$

If we put $\tau = \sigma_{i+1}\cdots\sigma_{j-2}\sigma_{j-1}$, then $(i,j) = \tau^{-1}\sigma_i\tau = \sigma_i^\tau$ and the result follows by inspection.

The transpositions of adjacent numbers $\sigma_k = (k, k+1)$ mentioned in Example 3.8 will come in very useful in the near future. We will call them *elementary transpositions*.

We illustrate this example in Figure 3.3, which shows how the transposition $(3, 8)$ can be expressed as a product of elementary transpositions. A similar argument shows that any transposition can be expressed as a product of elementary transpositions.

Definition 3.9

Suppose that $\omega_1, \omega_2 \in$ Sym Ω. We say that the permutations ω_1, ω_2 are *disjoint* if supp $\omega_1 \cap$ supp $\omega_2 = \emptyset$.

It is clear that disjoint permutations commute.

Proposition 3.10

Suppose that $\sigma \in S_n$, then σ is a product of disjoint cycles.

Proof

We deem the empty product of cycles to be the identity permutation. Suppose that $\sigma \neq$ id. The orbits of $\langle \sigma \rangle$ are disjoint. Restricting the domain of σ to a given non-trivial orbit yields a cycle. These cycles are disjoint and there are only finitely many of them since n is a natural number. The product of these finitely many cycles is σ since, considered as maps, σ and the product agree at every argument.

$\square$

Given any expression as a product of disjoint cycles where every $i \in \{1, 2, \ldots, n\}$ appears exactly once in a cycle, the supports of the non-trivial cycles are the non-trivial orbits of $\langle \sigma \rangle$ so there is no doubt about what they are. The cycles can be written in any order (a modest failure of uniqueness). There is also a little ambiguity in how each cycle is *written* $((1, 2, 3) = (2, 3, 1))$ but there is no doubt about what the cycles are.

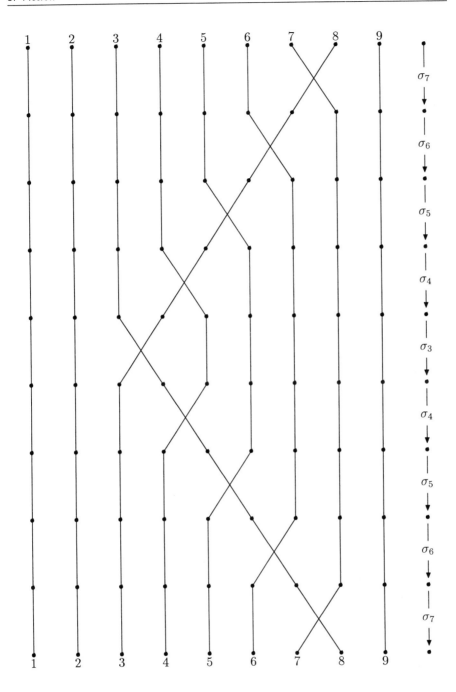

Figure 3.3 $(3, 8)$ as a product of elementary transpositions
(read from top to bottom)

Corollary 3.11

The expression of $\sigma \in S_n$ as a product of disjoint cycles is essentially unique.

Remark 3.12

We could work with elements of the symmetric group on an infinite set. Some but not all of the theory goes through without significant change. Every permutation is still a "product of cycles" but some cycles may be infinite, and moreover you may need an infinite number of cycles in the "product". Thus this product is a formal one, and is not taking place in the group. For example, in Sym ($\mathbb{N}$) there is the permutation

$$\prod_{i=1}^{\infty}(2i-1,2i).$$

You cannot multiply all those transpositions together in the group, but the meaning of the formal product is that even numbers swap with their predecessors.

Example 3.13

Consider $\nu \in S_6$ defined by

$$(1)\nu = 3, \ (2)\nu = 4, \ (3)\nu = 1, \ (4)\nu = 5, \ (5)\nu = 2, \ (6)\nu = 6.$$

The orbits of $\langle \nu \rangle$ are $\{1,3\}$, $\{2,4,5\}$ and $\{6\}$. The restriction of ν to each of these orbits is $(1,3)$, $(2,4,5)$ and the identity map (respectively). We have $\nu = (1,3)(2,4,5) = (2,4,5)(1,3)$.

Notice that $(1,2,3,\ldots,m) = (1,2)(1,3)\cdots(1,m)$ so this m-cycle is a product of transpositions, and the same remark applies to any m-cycle. It follows that any element of S_n is a product of transpositions. This is an important observation so we make it formally. Now Example 3.8 shows that any element of S_n is a product of elementary transpositions.

Proposition 3.14

The group S_n is generated by transpositions. In fact S_n is generated by the elementary transpositions.

Proof

Every permutation is a product of cycles, every cycle is a product of transpositions, and every transposition is a product of elementary transpositions.

□

Informally, we have shown that if you have n people in a row, and you wish to sort them into some order, then this may be accomplished by a sequence of moves, each of which consists of exchanging an adjacent pair of people.

Remark 3.15

Note that the definition of a cycle ensures that it has non-trivial support. However, it is sometimes convenient to think of the identity permutation as a cycle of length 1. For example, in S_6 we have the permutation $(1,2)(3,6)$. The fact that 4 and 5 are omitted from this notation can be a nuisance. We interpret (4) as the identity permutation since it sends 4 to 4 and leaves everything else fixed. Similarly (5) is the identity permutation, and therefore

$$(1,2)(3,6) = (1,2)(3,6)(4)(5)$$

and every number is mentioned. This swivel of the hips can sometimes be very convenient.

EXERCISES

3.3 Suppose that $\sigma \in S_n$. Write σ as a product of disjoint cycles. Show that the order of σ is the least common multiple of the lengths of these cycles (and include enough 1-cycles so that every element of $\{1, 2, \ldots, n\}$ is mentioned in the disjoint cycle representation of σ).

3.4 Suppose that Ω is a set of size 13. What is the smallest number n which does not arise as the order of an element of Sym Ω? Justify your answer.

3.5 Exhibit an element of Sym $\mathbb{Z}$ which is a formal infinite product of infinitely many disjoint infinite cycles (together with as many "1-cycles" in the product as you please).

3.2 Conjugacy in the Symmetric Group

Suppose that $(a_1, a_2, \ldots, a_t)$ is a cycle in S_n, and that $g \in S_n$. A direct calculation yields that

$$(a_1, a_2, \ldots, a_t)^g = g^{-1}(a_1, a_2, \ldots, a_t)g = ((a_1)g, (a_2)g, \ldots, (a_t)g).$$

For example

$$(1, 2, 3, 4, 5)^{(1,3,2,6)} = (3, 6, 2, 4, 5) = (2, 4, 5, 3, 6).$$

Since conjugation by g is an automorphism of S_n, it follows that we can conjugate a product of disjoint cycles in similar fashion. Thus conjugation in S_n is an extremely straightforward process. Moreover, if $\sigma \in S_n$ is represented as a product of disjoint cycles (including the "1-cycles") of lengths $u_1, u_2, \ldots, u_m$ (where $u_i \geq u_{i+1}$ for every $i < m$ and $\sum_j u_j = n$) then the same is true of every conjugate of σ. For example to work out $((1,2)(3,4,5))^{(1,2)(5,6)}$ we may assume that we are working in S_6. Now expressed in terms of cycles, $(1,2)(3,4,5) = (1,2)(3,4,5)(6)$ so the calculation is

$$((1,2)(3,4,5)(6))^{(1,2)(5,6)} = (2,1)(3,4,6)(5) = (1,2)(3,4,6).$$

Thus both $(1,2)(3,4,5)$ and its conjugate are products of disjoint cycles of lengths $3, 2$ and 1.

It gets better: the argument is reversible, so any two permutations which enjoy the same cycle shape are conjugate. When we write *cycle shape*, we mean the collection of lengths of cycles in an expression of an element as a product of disjoint cycles. In S_6 the permutation $(1,2)(4,5)$ has cycle shape 2,2,1,1. According to our assertion, it must be conjugate in S_6 to $(1,6)(3,4)$. The conjugating permutation $(2,6)(3,5,4)$ does the trick as we can calculate

$$((1,2)(4,5))^{(2,6)(3,5,4)} = (1,6)(3,4).$$

This indicates how the reverse argument goes in general. Suppose that $\alpha, \beta \in S_n$ have the same cycle shapes. We may express each of them as a product of disjoint cycles (including any cycles of length 1) with the lengths non-increasing. Thus

$$\alpha = (a_{11}, a_{12}, \ldots, a_{1n_1})(a_{21}, a_{22}, \ldots, a_{2n_2}) \cdots (a_{k1}, a_{k2}, \ldots, a_{kn_k})$$

and

$$\beta = (b_{11}, b_{12}, \ldots, b_{1n_1})(b_{21}, b_{22}, \ldots, b_{2n_2}) \cdots (b_{k1}, b_{k2}, \ldots, b_{kn_k}).$$

The required conjugating permutation γ such that $\alpha^\gamma = \beta$ is defined by $a_{ij} \mapsto b_{ij}$ for all legal pairs i, j.

We conclude that the conjugacy classes of S_n are parameterized (listed) by the possible cycle shapes. For example, there are three conjugacy classes in S_3 because $3, 2+1, 1+1+1$ describe the three possible cycle shapes. In S_5 there are 7 conjugacy classes because $5, 4+1, 3+2, 3+1+1, 2+2+1, 2+1+1+1$ and $1+1+1+1+1$ describe the 7 possible cycle shapes.

Remark 3.16

In particular, all transpositions in S_n are conjugate.

Definition 3.17

A permutation $\sigma \in S_n$ is said to be *even* if it is expressible as a product of an even number of transpositions. A permutation $\sigma \in S_n$ said to be *odd* if it is expressible as a product of odd number of transpositions. The empty product is the identity permutation and this is an even permutation since 0 is even.

When $n \geq 2$ there is no need for special pleading on behalf of the identity, since id $= (1,2)(1,2)$.

Example 3.18

$(1,2,3,4)(5,6) = (1,2)(1,3)(1,4)(5,6)$ is an even permutation.

Note that the inverse of an even permutation is even, and that the inverse of an odd permutation is odd. This is because the inverse of a product of transpositions is obtained by writing the transpositions in reverse order. For example we have

$$(1,2)(3,4)(4,5) \cdot (4,5)(3,4)(1,2) = \text{id}$$

and

$$(4,5)(3,4)(1,2) \cdot (1,2)(3,4)(4,5) = \text{id}.$$

It is clear that the even permutations form a subgroup of S_n. This subgroup A_n is called the alternating group. If $\tau \notin A_n$ and $n \geq 2$, then $\tau(1,2) \in A_n$ is an even permutation, so $\tau = \tau(1,2)(1,2) \in A_n(1,2)$ and therefore $S_n \subseteq A_n \cup A_n(1,2)$ and thus $S_n = A_n \cup A_n(1,2)$. We conclude that $|S_n : A_n| \leq 2$ when $n \geq 2$. In fact the index is 2, but we must eliminate the possibility that the index is 1. Both S_1 and A_1 are the trivial group, so $|S_1 : A_1| = 1$.

Remark 3.19

We must show that $A_n \neq S_n$ for $n \geq 2$, or equivalently, that $(1,2)$ is not an even permutation. It suffices to prove the following result.

Proposition 3.20

No element of S_n is both even and odd.

Proof

Suppose, for contradiction, that $\mu \in S_n$ is both even and odd. Express μ as a product of an odd number of transposition; writing these transpositions in reverse order yields μ^{-1} so μ^{-1} is an odd permutation. Thus the identity permutation $\mu\mu^{-1}$ is the product of an even permutation and an odd permutation and so is odd. By the discussion following Example 3.8, every transposition is a product of an odd number of elementary transpositions $\sigma_i = (i, i+1)$. We deduce that the identity permutation is the product of elementary transpositions

$$\mathrm{id} = \sigma_{i_1} \sigma_{i_2} \cdots \sigma_{i_m} \tag{3.2}$$

where m is an odd number. Now consider objects $a_1, \ldots, a_n$ being permuted, initially located in respective positions $1, 2, \ldots, n$. Each σ_{i_j} swaps the positions of two currently adjacent objects and does nothing else. Since the product (3.2) is the identity permutation, each pair of distinct objects must be swapped an even number of times, and so m must be even.

$$\square$$

Another way to view the final step of that argument is to consider Figures 3.4 and 3.5. Each pair of lines must cross an even number of times, so the total number of crossings must be even.

We conclude that A_n has index 2 in S_n for $n \geq 2$, and so $|A_n| = n!/2$ for these values of n. The group A_5 is particularly interesting for various reasons. It is the smallest non-abelian finite simple group, and it arises as the group of rigid symmetries of two of the five Platonic solids, the icosahedron and the dodecahedron. We will study this group at greater length in Chapter 4.

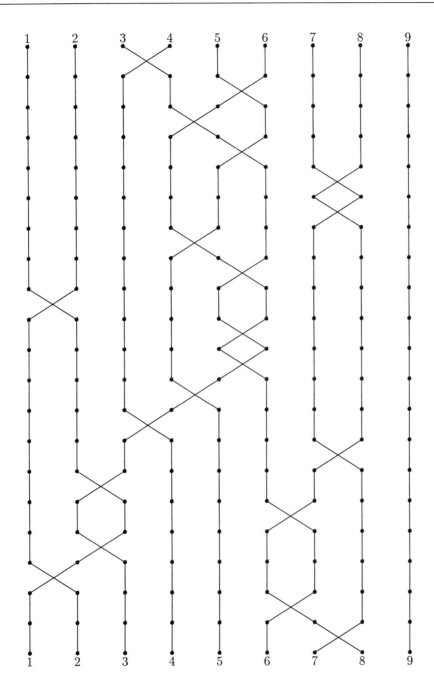

Figure 3.4 The identity permutation: the big picture

Each path from k to k crosses each other path an even number of times

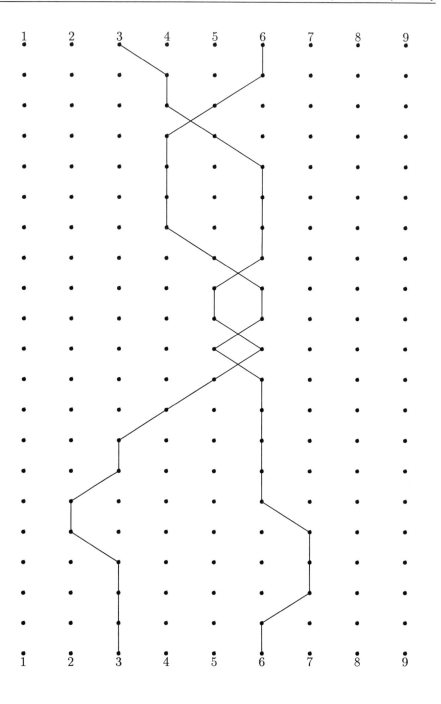

Figure 3.5 The identity permutation: two paths entwined

We omit all paths save those from 3 to 3 and 6 to 6. They cross 4 times

3.3 Group Actions

Suppose that Ω is a set. Recall that Sym Ω is the group of all bijections from Ω to Ω with composition of functions as the group operation. Now suppose that we have a homomorphism α from a group G to Sym Ω. This gives us a way of "multiplying" an element of Ω by an element of G to produce an element of Ω, via the rule ω times g is $(\omega)((g)\alpha)$. Thus we apply the bijection $(g)\alpha$ to an element ω of its domain, and this yields an element of Ω. If we denote this multiplication by a *dot*, we have two properties which are easy to check:

(i) $\omega \cdot 1 = \omega$ for all $\omega \in \Omega$.

(ii) $(\omega \cdot g) \cdot h = \omega \cdot (gh)$ for all $\omega \in \Omega$, and for all $g, h \in G$.

We now cut out the homomorphism.

Definition 3.21

We consider a map written as a dot. We say that the map $\cdot : \Omega \times G \to \Omega$ is an *action of G on Ω* if the laws (i) and (ii) hold.

We write this dot between its arguments, just as the map $+$ is written. After all, as any child knows, $+ : \mathbb{Z} \times \mathbb{Z} \to \mathbb{Z}$ and as an alternative to $((2,3))+ = 5$, you may have seen the notation $2 + 3 = 5$.

Remark 3.22

If we are given an action of G on Ω, then we can define a homomorphism $\beta : G \to$ Sym Ω via $(g)\beta : \omega \mapsto w \cdot g$ for every $\omega \in \Omega$ and every $g \in G$. If you start with the homomorphism, use it to define an action, and then use that action to define a homomorphism, then you recover the homomorphism you first thought of. Conversely, if you start with an action, a similar story unfolds. The dedicated reader should check these facts. Note that Sym Ω acts on Ω via $\omega \cdot \sigma = (\omega)\sigma$. Thus this business of a group acting on a set generalizes the notion of a permutation group.

Definition 3.23

If the group G acts on Ω then the *kernel of the action* is

$$K = \{g \mid \omega \cdot g = \omega \ \forall \omega \in \Omega\}.$$

This is the same subgroup of G as the kernel of the induced homomorphism $\beta : G \to$ Sym Ω, and so $K \trianglelefteq G$. An action is *faithful* if the kernel of the action is trivial. In this case the induced homomorphism is a monomorphism.

When studying such a group action, it is sometimes convenient to replace G by Im $\beta \simeq G/K$ because Im β is a permutation group on Ω.

We now revisit the alternating group A_n where $n \geq 2$. Recall that we had to do a little work to show that the even permutations in S_n formed a subgroup of index 2. We now have a way of doing this rather neatly. Let $\mathbb{Z}[X_1, X_2, \ldots, X_n]$ be the collection (ring) of all polynomials in n commuting variables with coefficients in $\mathbb{Z}$. Define an action of S_n on this ring via

$$f(X_1, X_2, \ldots, X_n) \cdot \sigma = f(X_{(1)\sigma}, X_{(2)\sigma}, \ldots, X_{(n)\sigma}).$$

Please check that this is an action! Moreover, this action has the additional multiplicative property that if $f_1, f_2 \in \mathbb{Z}[X_1, X_2, \ldots, X_n]$ and $\sigma \in S_n$, then

$$(f_1 f_2) \cdot \sigma = (f_1 \cdot \sigma)(f_2 \cdot \sigma).$$

Let $h = \prod_{i<j}(X_j - X_i)$. Thus, for example, if $n = 3$, then

$$h = (X_3 - X_1)(X_2 - X_1)(X_3 - X_2).$$

Back at the general case, note that for any $\sigma \in S_n$, as i, j range over $1 \leq i < j \leq n$, the polynomials $(X_{(j)\sigma} - X_{(i)\sigma})$ range over the polynomials $X_j - X_i$ where $1 \leq i < j \leq n$ save that some of these linear polynomials may have incorrect sign.

We use our example polynomial h when $n = 3$ and let $\sigma = (1, 2)$. Now

$$(X_3 - X_1)(X_2 - X_1)(X_3 - X_2) \cdot (1, 2) = (X_3 - X_2)(X_1 - X_2)(X_3 - X_1)$$

$$= (-1)(X_3 - X_1)(X_2 - X_1)(X_3 - X_2) = -h.$$

In general a factor of $(-1)^m$ is introduced where m counts the number of pairs $1 \leq i < j \leq n$ such that $(j)\sigma < (i)\sigma$.

Let $\Omega = \{h, -h\}$, so Ω is invariant under the action of S_n, and thus G acts on Ω. Note that the induced permutation group on Ω must be cyclic of order 2, and the kernel of the action K is a normal subgroup of index 2 in S_n. Now we have observed that $(1, 2) \notin K$. By Remark 3.16 all transpositions are conjugate in S_n. If just one of them were an element of K then they would all have to be elements of K since $K \trianglelefteq S_n$. Therefore $K\tau = K(1, 2) \neq K$ for each transposition τ. If σ is an even permutation it is the product of an even number of transpositions τ_i so

$$\sigma = \tau_1 \tau_2 \cdots \tau_{2m}.$$

Therefore

$$\begin{aligned}
K\sigma &= K\tau_1\tau_2\cdots\tau_{2m} = K\tau_1 \cdot K\tau_2 \cdots K\tau_{2m} \\
&= \left(K(1,2)\right)^{2m} = K(1,2)^{2m} = K.
\end{aligned}$$

Therefore every even permutation is an element of K and therefore $|S_n : A_n| = |S_n : K| \cdot |K : A_n| \geq 2$. This gives an alternative way to deal with the problem raised in Remark 3.19.

Incidentally, we have proved the following (given that a subgroup of index 2 must be normal).

Proposition 3.24

Suppose that $K < S_n$ and $|S_n : K| = 2$, then $K = A_n$.

This action of S_n on $\mathbb{Z}[X_1, X_2, \ldots, X_n]$ is very important. A polynomial $f \in \mathbb{Z}[X_1, X_2, \ldots, X_n]$ is called *symmetric* if it fixed by every $\sigma \in S_n$. Such polynomials comprise a subset (a *subring* if you know what that means) of $\mathbb{Z}[X_1, X_2, \ldots, X_n]$ which was first intensively studied by Sir Isaac Newton.

Example 3.25

We return to Example 1.16. Recall that D is the group of rigid motions of a cube. Let $\Lambda = \{1, 2, 3, 4\}$ be labels for the long diagonals. Now each $d \in D$ sends long diagonals to long diagonals, and this yields an action of D on Λ. In turn this means that there is an induced homomorphism $\lambda : D \to S_4$. The kernel of λ is trivial for geometrical reasons (pick up a cube, look and think) so λ is injective by Proposition 2.18(ii). However, D and S_4 both have size 24, so λ is surjective and therefore an isomorphism. We have finally justified the assertion made in Chapter 1 that D is a disguised version of S_4.

3.4 Orbits Form Partitions

Suppose that G acts on Ω. We define a relation $\sim$ on Ω via $\omega_1 \sim \omega_2$ if and only if there is $g \in G$ such that $\omega_1 \cdot g = \omega_2$. We now verify that this is an equivalence relation. If $\omega \in \Omega$, then $\omega \cdot 1 = \omega$ by the axioms, so $\omega \sim \omega$ for all $\omega \in \Omega$. Reflexivity is now established. Next suppose that $\omega_1, \omega_2 \in \Omega$ and $\omega_1 \sim \omega_2$. Therefore there is $h \in G$ such that $\omega_1 \cdot h = \omega_2$. Thus $(\omega_1 \cdot h) \cdot h^{-1} = \omega_2 \cdot h^{-1}$. We conclude that $\omega_1 = \omega_1 \cdot 1 = \omega_1 \cdot (hh^{-1}) = (\omega_1 \cdot h) \cdot h^{-1} = \omega_2 \cdot h^{-1}$. Thus symmetry is established. Finally suppose that $\omega_1 \sim \omega_2$ and $\omega_2 \sim \omega_3$. Thus there are $x, y \in G$ such that $\omega_1 \cdot x = \omega_2$ and $\omega_2 \cdot y = \omega_3$. Thus $\omega_1 \cdot (xy) = (\omega_1 \cdot x) \cdot y = \omega_2 \cdot y = \omega_3$. Thus transitivity is establishes and $\sim$ is an equivalence relation.

The equivalence classes of $\sim$ are called G-orbits, and the G-orbit containing $\omega \in \Omega$ is written ωG. For each $\omega \in \Omega$, let $G_\omega = \{g \mid g \in G, \omega \cdot g = \omega\}$. It is the work of a moment to verify that G_ω is actually a subgroup of G. The subgroup G_ω has various names. Some call it the *stabilizer* of ω, and write it as $\mathrm{Stab}_G(\omega)$. We will risk the more economical G_ω notation. Some people call G_ω the *isotropy group* of ω.

Example 3.26

Here are a collection of natural group actions.

(i) $\Omega = G$, and the action $\omega \cdot g = \omega g$.

(ii) $\Omega = G$, and the action $\omega \cdot g = g^{-1}\omega$.

(iii) $\Omega = G$, and the action $\omega \cdot g = g^{-1}\omega g$.

(iv) $H \leq G$, $\Omega = H\backslash G$, and the action $Hx \cdot g = Hxg$.

(v) $H \leq G$, $\Omega = \{H^x \mid x \in G\}$, and the action $H^x \cdot g = (H^x)^g = H^{xg}$.

(vi) Suppose that G acts on both A and B. Let $\Omega = \{f \mid f : A \to B\}$. Here are three natural actions of G on Ω

    (a) $f \cdot g : a \mapsto (a)f \cdot g$ for every $a \in A$.

    (b) $f \cdot g : a \mapsto (a \cdot g^{-1})f$ for every $a \in A$.

    (c) $f \cdot g : a \mapsto (a \cdot g^{-1})f \cdot g$ for every $a \in A$.

(vii) G acts on Γ and $\Omega = P(\Gamma)$, the power set of Ω. The action is defined as follows. If $S \subseteq \Gamma$, then $A \cdot g = \{ag \mid a \in A\}$.

Please verify that these alleged actions satisfy conditions (i) and (ii) of Definition 3.21.

Remark 3.27

There is an entirely analogous notion of a group G acting on a set Ω on the left. Everything turns round. For example, we can vary Example 3.26 (iii) by defining a left action of G on G via $g \cdot \omega = g\omega g^{-1}$. We must be careful here. You cannot define a right action by $\omega \cdot g = g\omega g^{-1}$ for all $g \in G$ and for all $\omega \in \Omega = G$. This is because if $x, y \in G$ and $\nu \in \Omega$ we have $(\nu \cdot x) \cdot y = yx\nu x^{-1}y^{-1}$ whereas $\nu \cdot (xy) = xy\nu y^{-1}x^{-1}$ and there is no guarantee that $yx\nu x^{-1}y^{-1} = xy\nu y^{-1}x^{-1}$. Law (ii) of Definition 3.21 is therefore not automatically satisfied.

However, if we are given a right action of G on Ω denoted by a dot, we can always manufacture a left action $*$ as follows. Define $g * \omega = \omega \cdot g^{-1}$ for each $\omega \in \Omega$ and $g \in G$. Certainly $1 * \omega = \omega \cdot 1 = \omega$ for every $\omega \in \Omega$. Also if $x, y \in G$ we have

$$(xy) * \omega = \omega \cdot (y^{-1}x^{-1}) = (\omega \cdot y^{-1}) \cdot x^{-1}$$

$$= (y * \omega) \cdot x^{-1} = x * (y * \omega).$$

Similarly given a left action denoted by a dot, we can manufacture a right action $*$ as follows. Let $\omega * g = g^{-1} \cdot \omega$. We leave the details to the reader.

Any group G acts on $\Omega = G$ by right multiplication via $\omega \cdot g = \omega g$. Similarly it acts on the left via $g * \omega = g\omega$. In fact these two actions commute in the sense that if $x, y \in G$ and $\omega \in \Omega = G$ we have $(x * \omega) \cdot y = x * (\omega \cdot y)$ because $x(\omega y) = (x\omega)y$ holds in a group. Thus we see that the associative law can be viewed as a commutativity condition; multiplication in the group from the left and from the right must commute!

Henceforth we allow ourselves to omit the irritating dot or $*$ from an action when mood and circumstances permit.

We now prove Cayley's theorem that every group arises as a permutation group. Thus if G is a group, then there is a copy of G which is a subgroup of Sym Ω for a suitable set Ω.

Proposition 3.28 (Cayley)

Suppose that G is a group. Let $\Omega = G$. Now G acts on itself by multiplication from the right. For each $g \in G$ we define a map $\theta_g : \Omega \to \Omega$ by $(w)\theta_g = w \cdot g$ for every $w \in \Omega$. It follows that each θ_g is a bijection, and that the map $\theta : G \to \text{Sym } \Omega$ defined by $g \mapsto \theta_g$ is a monomorphism, so Im $\theta \simeq G$.

Proof

Each θ_g has two-sided inverse $\theta_{g^{-1}}$, and so is a bijection. If $x, y \in G$ and $w \in \Omega$, then

$$(w)(xy)\theta = w(xy) = (wx)y = (w)((x)\theta \circ (y)\theta)$$

so θ is a homomorphism. If $k \in \text{Ker } \theta$ then $wk = w$ for every $w \in \Omega = G$ so $k = 1$. Thus Ker θ is the trivial group and θ is a monomorphism.

Of course, if you are one of that resolute band of people so repelled by our nasty little algebraic habit of writing maps on the right that you work out the maps-on-the-left version of everything in this book, then you will need to let G act on itself by multiplication from the left in that proof.

Proposition 3.29

Suppose that a group G acts on a set Ω. If $\omega \in \Omega$, then there is a natural bijection ψ between the right cosets of G_ω, and the orbit ωG.

Proof

The plan is to describe the map ψ, check that it is a map, and then verify that it is a bijection. Suppose that $L \in G_\omega \backslash G$, so $L = G_\omega x$ for some $x \in G$. We let $(L)\psi = \omega \cdot x$. Certainly $\omega \cdot x \in \omega G$ which is comforting. There is a well-definedness problem however. If $y \in G$ it may be that $G_\omega x = L = G_\omega y$. We had better be sure that $\omega \cdot x = \omega \cdot y$, otherwise our definition will not make any sense. Now, when $G_\omega x = G_\omega y$ it follows that $x = hy$ for some $h \in G_\omega$. In this event

$$\omega \cdot x = \omega \cdot (hy) = (\omega \cdot h) \cdot y = \omega \cdot y$$

as required. Thus ψ is well-defined.

Next we show that ψ is injective. If $(L_1)\psi = (L_2)\psi$ for $L_1, L_2 \in G_\omega \backslash G$ with $L_i = G_\omega x_i$ $(i = 1, 2)$, then $\omega \cdot x_1 = \omega \cdot x_2$ so $(\omega \cdot x_1) \cdot x_2^{-1} = (\omega \cdot x_2) \cdot x_2^{-1}$. We conclude that $\omega \cdot (x_1 x_2^{-1}) = \omega \cdot 1 = \omega$. Now $x_1 x_2^{-1} \in G_\omega$ so $G_\omega x_1 x_2^{-1} = G_\omega$ and therefore $L_1 = G_\omega x_1 = G_\omega x_2 = L_2$ and we have established that ψ is injective.

The surjectivity of ψ is straightforward. If $\mu \in \omega G$, then there is $g \in G$ such that $\omega \cdot g = \mu$. Therefore $(G_\omega g)\psi = \omega \cdot g = \mu$. We are done.

$\square$

If it so happens that we are dealing with finite sets, then this bijection yields an equality of natural numbers. Moreover, the following result has stunning application to finite groups acting on finite sets.

Proposition 3.30

Suppose that the group G acts on the finite set Ω, with distinct orbits $\omega_i G$ for $1 \leq i \leq t$. It follows that $|\Omega| = \sum_{i=1}^t |G : G_{\omega_i}|$.

Proof

The orbits form a partition of Ω so $|\Omega| = \sum_{i=1}^t |\omega_i G|$, and moreover $|\omega_i G| = |G : G_{\omega_i}|$.

$\square$

There is also a neat way to count the number of orbits when both G and Ω happen to be finite. We need some notation. If $g \in G$, then we let $\mathrm{Fix}(g)$ denote

$\{\omega \mid \omega \cdot g = \omega\}$. The elements of $\text{Fix}(g)$ are called *fixed points* of g. In some books you may find the following result described as Burnside's lemma. Peter Neumann has demonstrated that this counting technique predates William Burnside.

Proposition 3.31 (not Burnside)

Suppose that the finite group G acts on the finite set Ω, and that there are exactly t distinct orbits of G on Ω, then

$$t = \frac{1}{|G|} \sum_{g \in G} |\text{Fix}(g)| \, .$$

Proof

Let the orbits be $\omega_1 G, \cdots, \omega_t G$. Consider the set

$$\Gamma = \{(\omega, g) \mid \omega \in \Omega, g \in G, \omega \cdot g = \omega\} \subseteq \Omega \times G.$$

We count Γ in two ways, and then equate the answers. Letting g range over G we obtain that

$$|\Gamma| = \sum_{g \in G} |\text{Fix}(g)|. \qquad (3.3)$$

However it is also true that

$$|\Gamma| = \sum_{\omega \in \Omega} |G_\omega| = \sum_{i=1}^{t} \left(\sum_{\omega \in \omega_i G} |G_\omega| \right).$$

Now, for each $\omega \in \omega_i G$ we have $|G : G_\omega| = |\omega_i G| = |G : G_{\omega_i}|$ so $|G_\omega| = |G_{\omega_i}|$. Thus $|\Gamma| = \sum_{i=1}^{t} (\sum_{\omega \in \omega_i G} |G_{\omega_i}|)$. However, the number of elements in the orbit $\omega_i G$ is $|G : G_{\omega_i}|$ so

$$|\Gamma| = \sum_{i=1}^{t} |G : G_{\omega_i}||G_{\omega_i}| = \sum_{i=1}^{t} |G| = t|G|. \qquad (3.4)$$

We have used Lagrange's theorem (Theorem 1.33) to obtain that the product of $|G_{\omega_i}|$ and $|G : G_{\omega_i}|$ is $|G|$. We equate the two expressions for $|\Gamma|$ in (3.3) and (3.4), divide by $|G|$ and we are done.

□

Proposition 3.31 has the interpretation that the number of orbits of G acting on Ω is the average size of the set of fixed points of the elements of G. For example, the symmetric group S_n acts on $\{1, 2, \ldots, n\}$ in a natural way. The action has a single orbit, so an average permutation fixes exactly one element of $\{1, 2, \ldots, n\}$ (strange but true; try some particular values of n).

3.5 Conjugacy Revisited

As we have already remarked, one way in which a group G acts is on itself via $x \cdot g = x^g = g^{-1}xg$; we say that x has been conjugated by g. In this case, the orbits are called *conjugacy classes*, and the stabilizer of $x \in G$ is $C_G(x) = \{y \mid y \in G, x^y = x\}$. Notice that $x^y = x$ if and only if $yx = xy$, so $C_G(x)$ is the set of elements of G which commute with x. The jargon for this subgroup is the *centralizer* of x in G. Notice that the conjugacy class of x is finite if and only if $|G : C_G(x)| < \infty$. In fact if $S \subseteq G$, then we let

$$C_G(S) = \bigcap_{s \in S} C_G(s)$$

and speak of the centralizer in G of S. Note that set theoretic casuistry or authorial diktat ensures that $C_G(\emptyset) = G$. In every case, $C_G(S)$ is the intersection of a non-empty collection of subgroups of G. Thus $C_G(S)$ is a subgroup of G.

The group $C_G(G)$ is so important that it merits a special name; it is the *centre* or *center* of G and is written $Z(G)$ as a careful reader will recall. Notice that the conjugacy class of c has size 1 if and only if $c \in Z(G)$. Choose representatives x_j $(1 \leq j \leq s)$ for the s conjugacy classes of G which have size larger than 1. We have two splendid formulas given by Propositions 3.30 and 3.31 which we may unleash when G is a finite group acting on itself by conjugation. The first of these equations becomes *the class equation*.

$$|G| = |Z(G)| + \sum_{j=1}^{s} |G : C_G(x_j)|. \tag{3.5}$$

The second equation tells us that the number of conjugacy classes of G is

$$\frac{1}{|G|} \sum_{g \in G} |C_G(g)|. \tag{3.6}$$

Thus the number of conjugacy classes of G is the average size of the centralizer of an element. Observe that this makes perfect sense if the finite group G is abelian.

3.6 Enumeration

Proposition 3.31 has endless applications to enumeration (counting) problems. We give one example, and then supply a collection of problems for the reader to tackle.

Example 3.32

The vertices of a regular pentagon may each be coloured black or white. In how many essentially different ways can a regular pentagon be decorated in this way?

To answer this question, we need to decide what "essentially different" means in this context. We decide that two colourings are not essentially different if a rigid motion of the pentagon will take one colouring to the other. We allow ourselves to turn the pentagon over, as well as to spin it about its centre. In this toy example, it is easy to work out the answer without the application of Proposition 3.31. There is the all black colouring, there is one colouring with just one white vertex, there are two colourings with two white vertices and three black (the white vertices can be adjacent or not). Exchanging the roles of black and white we quickly finish by noting that there are

$$1 + 1 + 2 + 2 + 1 + 1 = 8$$

essentially different colourings of the vertices using two colours. Now, as a confidence building measure, we use not Burnside's method.

The dihedral group of order 10 acts naturally on the set of all possible colourings of the vertices of the regular pentagon using black and white. There are 2^5 such colourings. There are 10 elements of the dihedral group. The identity element, rotations through $2\pi/5, 4\pi/5, 6\pi/5$ and $8\pi/5$ about the centre in the plane of the pentagon. Finally there are five rotations through π (reflections) about an axis of symmetry. The identity fixes 2^5 colourings, each of the four non-trivial rotations fixes just two colourings (all black and all white), and the five reflections each fix 2^3 colourings (the vertices split into three orbits of sizes $1, 2$ and 2 and each region must be coloured monochromatically). Thus not Burnside's argument tells us that the total number of essentially different colourings is

$$\frac{2^5 + 4 \times 2^1 + 5 \times 2^3}{10} = 8.$$

EXERCISES

3.6 A Eurostandard standard is a flag comprising 3 horizontal stripes, each of which has the same height. The two dimensions of a Eurostandard standard must be in the golden ratio, and the stripes must be the direction of the longer dimension. There are c Europermitted colours which may be used in the manufacture of standards, jacks, pennants and bunting. How many essentially different Eurostandard standards are there?

3.7 A pair of colourings are deemed not essentially different if one can be obtained from the other by a rigid motion in Euclidean 3-space.

(a) How many essentially different ways are there to paint the vertices of a cube using c colours?

(b) How many essentially different ways are there to paint the edges of a cube using c colours?

(c) How many essentially different ways are there to paint the faces of a cube using c colours?

(d) How many essentially different ways are there to paint the vertices of a regular tetrahedron using c colours?

(e) How many essentially different ways are there to paint the edges of a regular tetrahedron using c colours?

(f) How many essentially different ways are there to paint the faces of a regular tetrahedron using c colours?

(g) How many essentially different ways are there to paint the vertices of a regular dodecahedron using c colours?

(h) How many essentially different ways are there to paint the edges of a regular dodecahedron using c colours?

(i) How many essentially different ways are there to paint the faces of a regular dodecahedron using c colours?

3.8 Let G be a finite group which has t conjugacy classes. Choose $x \in G$ uniformly at random. Choose $y \in G$ uniformly at random (with replacement, so that it is quite possible that $x = y$).

(a) What is the probability that x and y commute?

(b) What is the probability that x and y are conjugate?

3.9 (a) Let G be a finite group with normal subgroup N of index n. Show that the number of conjugacy classes of G is no more than n times the number of conjugacy classes of G which happen to be contained in N.

(b) Let G be a finite group with subgroup H of index n. Show that the number of conjugacy classes of G is no more than n times the number of conjugacy classes of G which intersect H non-trivially.

(c) What is the relationship between the results of the previous two parts of this question?

3.10 Justify the equivalence of group actions and the induced homomorphisms which is asserted in Remark 3.22.

3.7 Group Theoretic Consequences

Theorem 3.33 (Cauchy)

Suppose that G is a finite group and that p is a prime number which divides $|G|$. It follows that there is an element $g \in G$ of order p.

Proof

We induct on $|G|$, the result being vacuously true for the trivial group. Thus we may assume that G is not the trivial group and every proper subgroup of G has order coprime to p, else we are home by induction. The class equation (3.5) tells us that the centre $Z(G)$ has order divisible by p, so $G = Z(G)$ is abelian. If A, B are distinct maximal (proper) subgroups of G, then AB is a group (since G is abelian), so by maximality $G = AB$. Now $|G| = |AB| = |A||B|/|A \cap B|$ by Corollary 2.31. However, since $|A|$ and $|B|$ are coprime to p and $|A \cap B|$ is a factor of $|A|$, it follows that $|G|$ is coprime to p which is absurd. Therefore the non-trivial abelian group G has a unique maximal subgroup M. Since M is the only maximal subgroup of G, all proper subgroups of G will be contained in M. Choose $g \in G$ with $g \notin M$, then $\langle g \rangle = G$ since otherwise $\langle g \rangle \leq M$. Thus G is cyclic, and the result holds for cyclic groups by Proposition 1.45(ii).

$\square$

If you have not seen Cauchy's theorem before, I hope that you find the result rather striking. In a way it is a partial converse to Lagrange's Theorem, since that result tells us that the order of each element of a finite group G must divide $|G|$. Note that the full converse of Lagrange's theorem is false, since the group D of Example 1.16 has no element of order 6.

The following cute result follows quickly from Cauchy's theorem. The reader may wish to refresh her memory by looking at Definition 1.48 and the material on the dihedral groups which follows it.

Proposition 3.34

Let p be an odd prime. If a group G has order $2p$, then G is cyclic or dihedral.

Proof

By Cauchy's theorem G contains elements x of order 2 and y of order p. If $x^y \neq x$ then G contains two distinct involutions and $G = \langle x, x^y \rangle$ by Lagrange's theorem. Thus G is dihedral by Definition 1.48.

There remains the possibility that $x^y = x$, in which case x and y commute. However $G = \langle x, y \rangle$ by Lagrange's theorem so G is abelian. However an abelian group of order $2p$ must be cyclic by the structure theorem (Theorem 2.44).

$\square$

Let $S = P(G)$ be the set of subsets of a group G. We let G act on S via $X \cdot g = g^{-1} X g$. The orbit of X is sometimes written X^G, a collection of subsets of G in natural bijective correspondence with the cosets $G_X \backslash G$ where G_X is the set of all $y \in G$ such that $y^{-1} X y = X$, or if you prefer, the set of all $y \in G$ such that $Xy = yX$. This group G_X is called the *normalizer in G of X*, written $N_G(X)$. Recall that we introduced the centralizer $C_G(X)$ in Section 3.5. Notice that $C_G(X) \leq N_G(X)$, and that $C_G(x) = N_G(\{x\})$ if $x \in G$. We can squeeze out a little more. It is even true that $C_G(X) \trianglelefteq N_G(X)$. To check this, we take any $c \in C_G(X)$ and any $n \in N_G(X)$, and try to show that $c^n \in C_G(X)$. For any $x \in X$ we have

$$xc^n = xn^{-1}cn = n^{-1}(nxn^{-1})cn = n^{-1}c(nxn^{-1})n = c^n x$$

so $c^n \in C_G(X)$. That looks a little like magic. The proof is correct, but somehow the story has not been told. This is always an incentive to look a little deeper. Incidentally (just to jolt any complacent reader), perhaps X is empty! In that case $C_G(\emptyset) = N_G(\emptyset) = G$ and not a lot is going on.

The group $N_G(X)$ acts on X via $x \cdot n = n^{-1}xn$. This induces a group homomorphism $\psi : N_G(X) \to \operatorname{Sym} X$ in the usual way, and $\operatorname{Ker} \psi = C_G(X)$. This is a much better reason why $C_G(X) \trianglelefteq N_G(X)$. Note that if X is finite then $\operatorname{Sym} X$ is finite and therefore

$$|N_G(X) : C_G(X)| = |N_G(X)/C_G(X)| \leq |\operatorname{Sym} X| \text{ is finite.}$$

The non-strict inequality is a consequence of the First Isomorphism Theorem, since $N_G(X)/C_G(X) \simeq \operatorname{Im} \psi \leq \operatorname{Sym} X$.

It is worth investigating what happens when X is not just an arbitrary subset of G, but is special. Perhaps you are wondering what happens when X is a subgroup of G. In this case $X \leq N_G(X)$ and $Z(X) = X \cap C_G(X)$. If X is actually abelian, we have $X \leq C_G(X)$. In Figure 3.6 we show both the general case, and the collapse ($X = X \cap C_G(X)$ and $XC_G(X) = C_G(X)$) which happens when X is abelian. One can think of the collapse as the two vertical sides of the parallelogram shrinking to nothing.

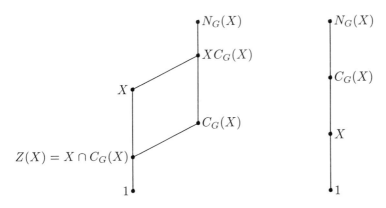

The general case   The case when X is abelian

Figure 3.6 X is a subgroup of G

There is another splendid result which flows from the theory of group actions which is very useful when studying simple groups.

Proposition 3.35 (Poincaré)

Suppose that G is a group containing a subgroup H of finite index n. Let $K = \cap_{x \in G} H^x$ so $K \leq H$. It follows that $m = |G : K|$ finite and $K \trianglelefteq G$. Moreover $n \mid m \mid n!$.

Proof

Let $\Omega = H\backslash G$ so Ω is a finite set of size n. Right multiplication is a natural transitive action of G on Ω. Therefore the stabilizer G_H has index n in G by Proposition 3.29, and since $H \leq G_H$ it follows that $H = G_H$. Notice also that for each $x \in H$ the stabilizer of Hx is H^x. We justify this as follows. Certainly H^x stabilizes Hx since for each $h \in H$ we have $Hxx^{-1}hx = Hx$. Conversely if y stabilizes Hx, then $Hxy = Hx$ so $xyx^{-1} \in H$ and therefore $y \in x^{-1}Hx = H^x$.

The kernel of the action is the intersection the stabilizers of each Hx, and since it does not matter if use extra copies of a stabilizer when forming our intersection, we deduce that $K = \cap_{x \in G} H^x \leq H$ (though we could get away with intersecting over all x in a right transversal of H). Moreover, by the first isomorphism theorem the image of the induced homomorphism from G to Sym Ω is isomorphic to G/K. Thus G/K is isomorphic to a subgroup of the symmetric group S_n and so has order dividing $n!$, but this means that $|G : K|$ divides $n!$. Also $|G : K| = |G : H||H : K|$ so $n \mid |G : K|$.

$\square$

Corollary 3.36

Suppose that G is a simple group and H is a subgroup of G with $1 < |G : H| = n < \infty$. It follows that $|G|$ divides $n!$ and so $|G| \leq n!$. It also follows that an infinite simple group has no proper subgroup of finite index.

Remark 3.37

Suppose that H is a subgroup of G. The set of all G-conjugates of H is invariant (closed) under conjugation by elements of G, so $\cap_{x \in G} H^x \trianglelefteq G$. We call this intersection $\mathrm{core}_G(H)$ the core of H in G, or just the core of H if the context makes G clear. Notice that if $K \trianglelefteq G$ and $K \leq H$, then $K = K^x \leq H^x$ for every $x \in G$, so $K \leq \mathrm{core}_G(H)$. Therefore $\mathrm{core}_G(H)$ is the largest normal subgroup of G which is contained in H. This presents us with a powerful technique. If we are in a position to know that there is no non-trivial normal subgroup of G contained in H, then we can deduce that $\mathrm{core}_G(H) = 1$.

As we argued in the proof of Poincaré's Proposition 3.35, we can obtain the core of H in G by intersecting all H^x as x ranges over a right transversal for H in G. If it so happens that $|G : H|$ is finite, then we are intersecting only finitely many subgroups, and moreover they all have the same finite index since conjugation by a particular group element is an automorphism of G. Now Proposition 1.32 applies, and we have another way of seeing that (in these circumstances) $|G : \mathrm{core}_G(H)|$ is finite.

3.8 Finite p-groups

Let p be a prime number. A group P is a *p-group* if the order of each element of P is a power of p. Infinite p-groups can be quite peculiar. It was Ol'shanskii and Rips who constructed the so-called *Tarski monsters*. These are infinite groups P where the only subgroups are the trivial group, the group P itself, and cyclic groups of order p. Anyone familiar with the rich complexity of the subgroup structure of finite p-groups should find this quite remarkable. For example, when p is a prime number, the group $\mathbb{Z}_p^3 = \mathbb{Z}_p \oplus \mathbb{Z}_p \oplus \mathbb{Z}_p$ has $p^2 + p + 1$ subgroups of order p, and the same number of subgroups of order p^2. Each subgroup of order p^2 contains $1 + p$ subgroups of order p, and any pair of distinct subgroups of order p^2 intersect in a subgroup of order p. This information is presented as fact but is not justified. The inquisitive reader may wish to fill in the details.

We focus on finite p-groups. If P is a group with $|P|$ a power of p, then Lagrange's theorem forces each subgroup of P to have p-power order. This includes all the cyclic subgroups, so each element of P has p-power order. On the other hand, if P is a finite p-group and q is a prime number which divides $|P|$, then Cauchy's Theorem 3.33 ensures that there is an element of P which has order q. Therefore $q = p$, so $|P| = p^n$. Thus to say that P is a finite p-group is the same as saying that P has order p^n for some $n \geq 0$.

Proposition 3.38

If P is a finite p-group and $P \neq 1$, then $Z(P) \neq 1$.

Proof

Use the class equation, where there are s conjugacy classes which are not singleton sets, and a representative x_j is selected from each one. Now

$$p^n = |P| = |Z(P)| + \sum_{j=1}^{s} |P : C_P(x_j)|,$$

so we see that p divides $|Z(P)|$. However, $1 \in Z(P)$ so $|Z(P)| \neq 0$. Thus $|Z(P)| \geq p$, and so $Z(P) \neq 1$.

$\square$

Corollary 3.39

If P is a p-group of order p^2, then P must be abelian. This is because if $Z(p)$ has order p, then we may choose $x \in P - Z(P)$. Now $x \in C_P(x)$ and $Z(P) \leq C_P(x)$ so $|C_P(x)| \geq p + 1$. By Lagrange's theorem $C_P(x) = P$ so $x \in Z(P)$ which is absurd. Therefore $Z(P) = P$ is abelian.

Proposition 3.40

Let P be a finite p-group, and suppose that $1 \leq H < P$. It follows that $H < N_P(H)$.

Proof

The conditions ensure that P is not the trivial group. Thus Proposition 3.38 applies and $Z(P)$ is also not the trivial group. If it so happens that $Z(P) \leq H < P$ then factoring out $Z(P)$ we see that $H/Z(P)$ is strictly contained in its normalizer in $P/Z(P)$ since the result holds by inductive hypothesis in the smaller group $P/Z(P)$. The Correspondence Principle applies so $H < N_P(H)$.

Thus we are done unless some $z \in Z(P)$ is not an element of H. However, in this event z commutes with every element of H and so $z \in N_P(H)$. Therefore $H < N_P(H)$.

Lemma 3.41

Suppose that G is a group with subgroups P, Q which are both finite p-groups. If $Q \leq N_G(P)$ then PQ is a p-subgroup of G.

Proof

Since Q normalizes P we have $P, Q \leq N_G(P)$ and $P \trianglelefteq N_G(P)$ so PQ is a group. Now $|PQ| = |P||Q|/|P \cap Q|$ is a power of p, and we are done. We have used Eq. (1.7) or if you prefer, Corollary 2.31.

$\square$

Corollary 3.42

If P is a largest possible finite p-subgroup of a group G, and Q is a finite p-subgroup of G which normalizes P, then $Q \leq P$.

Definition 3.43

Let G be a finite group of order $p^a n$ where p does not divide n. Any subgroup of G of order p^a is called a Sylow p-subgroup, and the set of all such subgroups is denoted $\mathrm{Syl}_p(G)$.

Theorem 3.44 (Sylow)

Let G be a finite group of order $p^a n$ where the prime p does not divide n.

(i) There is a Sylow p-subgroup P of G.

(ii) If $Q \leq G$ and Q is a p-group, then there is $T \in \mathrm{Syl}_p(G)$ such that $Q \leq T$.

(iii) All elements of $\mathrm{Syl}_p(G)$ are conjugate in G.

(iv) $|\mathrm{Syl}_p(G)|$ divides n, and is congruent to 1 modulo p.

Proof

We prove (i) by induction on $|G|$, the result being true for the trivial group. We may therefore assume that $|G| > 1$. The class equation tells us that $|G| =$

$|Z(G)| + \sum_{j=1}^{s} |G : C_G(x_j)|$ where the x_j are a collection of representatives for those conjugacy classes which are not singleton sets. Therefore each $C_G(x_j)$ is a proper subgroup of G. If $p \mid |G : C_G(x_j)|$ for every j, then $p \mid |Z(G)| \neq 1$. Thanks to Cauchy's Theorem 3.33 we can choose $z \in Z(G)$ of order p so $\langle z \rangle \trianglelefteq G$. There is the natural projection $\pi : G \to G/\langle z \rangle$. By induction, there is a Sylow p-subgroup P_1 of $G/\langle z \rangle$. This group has order p^{a-1} since $|G/\langle z \rangle| = p^{a-1}n$. The preimage of P_1 under π is $P \leq G$ where $P/\langle z \rangle$ has order $p^{a-1} = |P|/p$. Thus $|P| = p^a$ and we have found a Sylow p-subgroup of G.

The other possibility is that there is some x_j with p not a divisor of $|G : C_G(x_j)|$, so $|C_G(x_j)| = p^a m$ with $m < n$ and m coprime to p. By induction, $C_G(x_j)$ has a subgroup P of order p^a. Now $P \leq G$ so P is a Sylow p-subgroup of G.

Now for part (ii). Suppose that Q is a p-subgroup of G. We have a Sylow p-subgroup P of G. Let $\Omega = \{P^g \mid g \in G\}$ denote the set of G-conjugates of P. Now Q acts on Ω by conjugation. We have $|\Omega| = |G : N_G(P)| = \sum_i |\Omega_i|$ where the Ω_i are the Q-orbits. If we choose $P_i \in \Omega_i$, we have $|\Omega_i| = |Q : N_Q(P_i)|$ where we define $N_Q(P_i) = N_G(P_i) \cap Q$. Orbits therefore have size divisible by p unless Q normalizes P_i, in which case $Q \leq P_i$ by Corollary 3.42. Temporarily letting Q be P, we deduce that $|\Omega| \equiv 1 \bmod p$ since P then is contained in exactly one Sylow p-subgroup of G. Now revert to letting Q be an arbitrary p-subgroup of G, we deduce that some $|\Omega_i|$ is not divisible by p. Thus Q normalizes P_i and so $Q \leq P_i$ by Corollary 3.42.

Now for part (iii). We have already shown that $|\Omega| \equiv 1 \bmod p$. Now suppose (for contradiction) that $\Omega \neq \mathrm{Syl}_p(G)$, so that there is a Sylow p-subgroup T which is not conjugate to P. Let T act on Ω by conjugation. All T-orbits have size divisible by p, since otherwise T would have to normalize, and hence be contained in, and hence be, a conjugate of P. Since Ω is a disjoint union of sets of size divisible by p we deduce that $|\Omega| \equiv 0 \bmod p$. However, we already know that $|\Omega| \equiv 1 \bmod p$ so this is absurd. Thus $\Omega = \mathrm{Syl}_p(G)$, and all Sylow p-subgroups of G are conjugate.

We have already established the second part of (iv). Now G acts on $\mathrm{Syl}_p(G) = \Omega$ by conjugation, and there is just one orbit. The stabilizer of $P \in \mathrm{Syl}_p(G)$ is $N_G(P)$, so $|\mathrm{Syl}_p(G)| = |G : N_G(P)|$. Since $P \leq N_G(P)$ it follows that $p^a \mid |N_G(P)|$, and so $|G : N_G(P)|$ must divide n.

$\square$

Proposition 3.45

Suppose that $N \trianglelefteq G$, where G is a finite group. Suppose that $P \in \mathrm{Syl}_p(G)$, then $PN/N \in \mathrm{Syl}_p(G/N)$ and $P \cap N \in \mathrm{Syl}_p(N)$.

Proof

Indices multiply so $|G : PN||PN : P| = |G : P|$ is coprime to p, and so therefore is $|G : PN|$. The second isomorphism theorem tells us that $PN/N \simeq P/P \cap N$ has order a power of p. Now $|G/N : PN/N| = |G : PN|$ by the memorable version of the third isomorphism theorem (Theorem 2.29) so PN/N has p-power order but has index in G/N which is coprime to p. Therefore $PN/N \in \mathrm{Syl}_p(G/N)$.

Now $P \cap N$ is of p-power order since $P \cap N \leq P$. Also

$$|PN| = |P|(|N|/|P \cap N|) = |P||N : P \cap N|.$$

If p divides $|N : P \cap N|$, then $p|P|$ divides $|PN|$ which divides $|G|$. This is impossible since $P \in \mathrm{Syl}_p(G)$. Therefore $P \cap N$ has p-power order but has index in N which is coprime to p. Therefore $P \cap N \in \mathrm{Syl}_p(N)$.

$\square$

The fact that $N \cap P$ is a Sylow p-subgroup of $N \trianglelefteq G$ is very useful.

EXERCISES

3.11 Suppose that G is a finite group and $P \in \mathrm{Syl}_p(G)$. Show that if $x \in G$, then $N_G(P^x) = N_G(P)^x$.

3.12 Let G be a finite group. Suppose that for all $P, Q \in \mathrm{Syl}_p(G)$ either $P = Q$ or $P \cap Q = 1$. Show that G contains exactly

$$|G : N_G(P)|(|P| - 1)$$

non-identity elements of p-power order.

3.13 Suppose that G is a finite group with $N \trianglelefteq G$. Choose $Q \in \mathrm{Syl}_p(N)$. Show that $N_G(Q)N = G$. *This result (or rather its proof) is called the Frattini argument.*

3.14 Suppose that G is a finite group and that $P \in \mathrm{Syl}_p(G)$. Also suppose that $N_G(P) \leq T \leq G$. Prove that $N_G(T) = T$.

3.9 Multiple Transitivity and Primitivity

Definition 3.46

Suppose that a group G acts on a set Ω. We say that G acts transitively on Ω

if the action of G has a single orbit. Equivalently, whenever $\omega_1, \omega_2 \in \Omega$ then there exists $g \in G$ such that $\omega_1 g = \omega_2$.

Thus if $\omega \in \Omega$ and Ω is finite, then $|G : G_\omega| = |\Omega|$ if and only if G acts transitively on Ω.

Example 3.47

The natural action of $\langle (1, 2, 3, \ldots, n) \rangle$ on $\{1, 2, 3, \ldots, n\}$ is transitive.

Definition 3.48

Suppose that m is a natural number, and that a group G acts on a set Ω. We say that G acts *m-transitively on* Ω if whenever two ordered m-tuples $(a_1, a_2, \ldots, a_m)$ and $(b_1, b_2, \ldots, b_m)$ have entries drawn from Ω, and neither list contains a repeated element, then there is $g \in G$ such that $a_i g = b_i$ for every i in the range $1 \le i \le m$. Note that an entry may appear in both lists but it may not appear more than once in each of the lists separately.

Thus 1-transitivity is the same notion as transitivity, and the multiple transitivity that we have been discussing is a tightening of the transitivity condition. It is harder to be a 2-transitive action than just to be a transitive action, and harder still to be a 3-transitive action than just to be a 2-transitive action and so on.

Example 3.49

It is clear that the symmetric group S_n acts n-transitively in its natural action on $\{1, 2, \ldots, n\}$. Moreover if Λ is an infinite set, then Sym Λ acts m-transitively on Λ for every finite m. Looking back to Example 3.47, and that the action given there is not 2-transitive for $n \ge 3$.

There is an elegant way of saying that G acts m-transitively on Ω. We have an induced action of G on

$$\Omega^m = \underbrace{\Omega \times \Omega \times \cdots \times \Omega}_{m \text{ copies of } \Omega}$$

defined by

$$(\omega_1, \omega_2, \ldots, \omega_m) \cdot g = (\omega_1 g, \omega_2 g, \ldots, \omega_m g)$$

for all relevant m-tuples and for all $g \in G$.

Let $\overline{\Omega}^m$ denote the set of elements of Ω^m which do not have repeated entries. This set is invariant under the action of G (this is a posh way of saying that $x \cdot g \in \overline{\Omega}^m$ whenever $x \in \overline{\Omega}^m$ and $g \in G$). The statement that G acts m-transitively on Ω can be recast as the assertion that G acts transitively on $\overline{\Omega}^m$.

Let V be a finite dimensional vector space over $\mathbb{R}$, and let $\mathrm{GL}(V)$ be the general linear group on V discussed in Example 1.15. In what follows, we discuss natural actions (i.e. the obvious actions), and to avoid triviality, we will assume that $V \neq 0$. In order to be consistent with the philosophy of doing linear algebra on the left, we shall let $\mathrm{GL}(V)$ act naturally on various sets *on the left*. This causes no difficulty whatsoever. There is no hope that $\mathrm{GL}(V)$ acts transitively on V since the zero vector is a fixed point of every element of $\mathrm{GL}(V)$. However, $\mathrm{GL}(V)$ does act transitively on the non-zero vectors of V since any non-zero vector can be part of a basis, and any ordered basis can be sent to any other ordered basis by an element of $\mathrm{GL}(V)$. However, $\mathrm{GL}(V)$ does not act 2-transitively on the non-zero vectors of V, since if $v \neq 0$ then there is no element $g \in \mathrm{GL}(V)$ such that simultaneously $g \cdot v = 2v$ and $g \cdot 2v = v$.

This group $\mathrm{GL}(V)$ has a natural 2-transitive action on the set of 1-dimensional subspaces of V (and to avoid triviality, V had better be at least 2-dimensional). The reader in need of geometric support should think of the set of all straight lines through the origin in 3-dimensional space, a sort of infinite porcupinewith its point-sized body at the origin. The collection of 1-dimensional subspaces of a vector space V is called the projective space $P(V)$. The natural action of $\mathrm{GL}(V)$ on $P(V)$ induces a homomorphism from $\mathrm{GL}(V)$ to $\mathrm{Sym}\,P(V)$. The kernel of this map consists of the scalar linear transformations (why?) and the image of the induced homomorphism is called $\mathrm{PGL}(V)$, the projective general linear group.

EXERCISES

3.15 Justify the assertion of the previous paragraph that if V is a 3-dimensional $\mathbb{R}$-vector space, then the natural action of $\mathrm{GL}(V)$ on the set of 1-dimensional subspaces is 2-transitive.

3.16 Show that if V is an n-dimensional $\mathbb{R}$-vector space with $n \geq 4$, then the natural action of $GL(V)$ on the set of all 2-dimensional subspaces is not 2-transitive.

3.17 Show that when $n \geq 3$, the alternating group A_n has a natural $(n-2)$-transitive action on $\{1, 2, \ldots, n\}$.

Suppose that G acts on a set Ω and that $B \subset \Omega$. Recall that there is an induced action of G on the set $P(\Omega)$ of all subsets of Ω ($P(\Omega)$ is called the power set of Ω). The action is as follows: if $g \in G$ and $S \subset \Omega$, then $S \cdot g = \{gs \mid s \in S\}$.

Definition 3.50

Suppose that G acts on a set Ω and that $B \subset \Omega$. Suppose that $B \subseteq \Omega$ has the property that whenever $g \in G$ we have either $Bg = B$ or $Bg \cap B = \emptyset$. We say that B is a *block* of the action of G on Ω.

A block B is said to be *trivial* if B is empty, a singleton set or the whole of Ω. Such subsets of Ω are the obvious and unavoidable blocks.

Definition 3.51

Suppose that G acts on a set Ω. The action is said to be *primitive* if there are no non-trivial blocks. Otherwise we say that the action is *imprimitive*.

Proposition 3.52

If the group G acts 2-transitively on Ω, then the action is primitive.

Proof

Suppose for contradiction that B is a non-trivial block. Choose $a_1, a_2 \in B$ with $a_1 \neq a_2$. This is possible since $|B| \geq 2$. Choose $b_1 = a_1$ and $b_2 \in \Omega - B$. This is possible since B is a proper subset of Ω. Thus $b_1 \neq b_2$. By 2-transitivity there is $g \in G$ such that $a_1 g = b_1$ and $a_2 g = b_2$. Now $Bg \ni b_2 \notin B$ so $B \neq Bg$, but $b_1 \in B \cap Bg$ so $B \cap Bg \neq \emptyset$. This is absurd.

$\square$

Transitive actions which are primitive are particularly interesting. They are intimately connected with maximal subgroups.

Definition 3.53

Let G be a group. A proper subgroup M of G is *maximal* if whenever $M \leq H < G$, it follows that $M = H$.

There are groups which have no maximal subgroups, but non-trivial finite groups always have them. Indeed, if S is any proper subgroup of the non-trivial finite group G, then there is a maximal subgroup M with $S \leq S < G$.

Proposition 3.54

Suppose that G acts transitively on Ω, where $|\Omega| \geq 2$. The following are equivalent:

(a) The action is primitive.

(b) If $w \in \Omega$, then the stabilizer G_w is a maximal subgroup of G.

(c) For every $\omega \in \Omega$ the stabilizer G_ω is a maximal subgroup of G.

Proof

We first show that (b) and (c) are equivalent. Suppose that $\omega, \omega' \in \Omega$. By transitivity there is $x \in G$ such that $wx = w'$. Notice that $x^{-1}G_w x \subseteq G'_\omega$. Also $xG_{\omega'}x^{-1} \subseteq G_w$ so $G_{\omega'} \subseteq x^{-1}G_w x$. It follows that $x^{-1}G_w x = G_{\omega'}$. Conjugation by x is an automorphism of G, so G_w is a maximal subgroup of G if and only if $G_{\omega'}$ is a maximal subgroup of G. Thus conditions (b) and (c) are equivalent.

Now suppose that (a) holds. Suppose that $\omega \in \Omega$ and that $G_\omega \leq H \leq G$. By Proposition 3.29 there is a natural bijection between $G_\omega \backslash G$ and Ω defined by $G_\omega x \mapsto \omega x$. Let $B = \omega H$. We now verify that B is a block. Suppose that $g \in G$ and $B \cap Bg \neq \emptyset$. Thus there are $h, h' \in H$ with $\omega h g = \omega h'$. Therefore $hgh'^{-1} \in G_\omega \leq H$ so $g \in H$ and $Bg = \omega Hg = \omega H = B$. If $G_\omega < H$, then $|B| \geq 2$. Moreover if $H < G$, then $B \neq \Omega$ because of the bijection mentioned above. Since the action is primitive we deduce that $H = G_\omega$ or $H = G$, so (c) holds.

Finally we suppose that (c) holds. Suppose that B is a block for the action and that $|B| \geq 2$. Let $H = \{g \mid B \cdot g = B\}$. Pick $\omega \in B$. Now $G_\omega \leq H$ since B is a block. Choose $\omega' \in B - \{\omega\}$ which is possible since $|B| \geq 2$.

By transitivity there is $y \in G$ with $\omega y = \omega'$. Since B is a block we have $By = B$ but of course $y \notin G_\omega$. Thus H strictly contains G_ω. By maximality of G_ω we deduce that $H = G$. Now for every $g \in G$ we have $\omega g \in \omega H = B$. However, G acts transitively on Ω so $B = \Omega$. We conclude that the action of G on Ω is primitive, so (a) holds.

$\square$

4
Entertainments

We will now take the basic theory developed in the first part of this book, and play with it to show a little of what can be done with our theoretical machinery. This chapter is crudely divided into two parts concerning finite and infinite groups. This division is easy to make, but in fact is not really defensible. Some of the techniques for studying infinite groups consist of viewing them as limiting cases (or generalizations) of finite groups. On the other hand some geometric perspectives on group theory work just as well for finite as for infinite groups. Certain properties of finite groups have the aroma of their infinite relatives, whereas some of the strictly combinatorial (counting) arguments which flow from Sylow's theorem have a pungent finite odour.

The collection of all infinite groups is so rich and varied that it is not possible to say very much about all of them at once, save for the theory for all groups which we have developed already. There are two ways forward. One approach is to impose some conditions on the class of infinite groups under consideration. A second strategy is to develop techniques for attempting to study any particular group, so that if it suddenly becomes very important to know as much as possible about a group which has arisen naturally (possibly in topology or geometry), then this array of methods can be deployed in order to facilitate our understanding. We illustrate these two approaches in Chapters 5 and 6.

4.1 The Finite Case

We will quickly summarize some of the key results which we plan to use in the first part of this chapter. We make the blanket assumption that G is a finite group, that P is a Sylow p-subgroup of G, and that p is a prime divisor of $|G|$. This list is here just to serve as a reminder of some of what we now know.

(a) If G is an abelian simple group, then $|G|$ is a prime number and G is a cyclic group. [Proposition 2.27]

(b) If $H < G$, and $|G : H| = 2$, then $H \lhd G$. [Exercise 2.12, p.52]

(c) If $H, K \leq G$, then

$$|HK| = \frac{|H||K|}{|H \cap K|}$$

[Eq. (1.7) and Corollary 2.31]. The set HK is a subgroup of G precisely when $HK = KH$. This condition will hold if either K or H is normal in G, and more generally if $H \leq N_G(K)$ or $K \leq N_G(H)$ [Exercise 1.13, p.31].

(d) If $x \in G$, then $N_G(P^x) = N_G(P)^x$. Remember that conjugation by a fixed $x \in G$ is an automorphism (self-isomorphism) of G, and so this is what you would expect. [Exercise 3.10, p.120].

(e) Suppose that $P \in \mathrm{Syl}_p(G)$. We have $|\mathrm{Syl}_p(G)| = 1$ if and only if $P \lhd G$. This is because all Sylow p-subgroups of G are conjugate, and moreover every conjugate of P will be a Sylow p-subgroup [Theorem 3.44].

(f) Suppose that for all $P, Q \in \mathrm{Syl}_p(G)$ either $P = Q$ or $P \cap Q = 1$. It follows that G contains exactly $|G : N_G(P)|(|P| - 1)$ non-identity elements of p-power order [Exercise 3.11, p.120].

(g) Suppose that H is a proper subgroup of the finite simple group G. It follows that $|G|$ divides $|G : H|!$ (and so $|G| \leq |G : H|!$). This is Poincaré's argument [Proposition 3.35].

(h) If $P \in \mathrm{Syl}_p(G)$ and $Q \leq N_G(P)$ with $|Q|$ a power of p, then $Q \leq P$ [Lemma 3.41].

(i) If $|G| = p$ is a prime number, then G is a cyclic group [Proposition 1.37].

(j) If $|G| = p^2$ is the square of a prime number, then either $G \simeq C_{p^2}$ or $G \simeq C_p \times C_p$. There is no possibility that $C_p \times C_p \simeq C_{p^2}$ since the first group contains no element of order p^2. [Corollary 3.39 shows that G is abelian. The structure theorem for finite abelian groups (Theorem 2.44) then applies.]

(k) If $|G| = 2p$ where p is an odd prime number, then $G \simeq C_{2p}$ or $G \simeq D_{2p}$. There is no possibility that $C_{2p} \simeq D_{2p}$ since the second group is not abelian [Proposition 3.34].

We have assembled quite a tool-box now. We will illustrate how powerful these ideas can be by looking at groups of small order. There can be no technique more crude than working through the natural numbers one at a time, pausing to reason at each step.

Order 1: There is only the trivial group of order 1.

Order 2: Since 2 is a prime number, any group of order 2 is a copy of C_2.

Order 3: Since 3 is a prime number, any group of order 3 is a copy of C_3.

Order 4: Since $4 = 2^2$ is the square of a prime number, any group of order 4 must be either a copy of C_4 or of $C_2 \times C_2$.

Order 5: Since 5 is a prime number, any group of order 5 is a copy of C_5. As you can see, classifying groups of prime order is a bit of a doddle.

Order 6: Since $6 = 2 \cdot 3$ is twice an odd prime number, every group of order 6 must either be C_6 or D_6 thanks to the result (k). The latter group is not abelian, whereas C_6 is abelian. Notice that there are two other obvious groups of order 6 : the group $C_2 \times C_3$ and the group S_3. Since the first group is abelian and the second is not, we deduce that $C_2 \times C_3 \simeq C_6$ and $S_3 \simeq D_6$.

Order 7: Since 7 is a prime number, any group of order 7 is a copy of C_7.

Order 8: So far life has been kind to us. Now is the time when work actually begins. There are three different pairwise non-isomorphic abelian groups of order 8. They are $C_8, C_2 \times C_4$ and $C_2 \times C_2 \times C_2$. The structure theorem for finite abelian groups tells us that these three groups are pairwise non-isomorphic, but we can see it immediately. A group of type C_8 contains an element of order 8 but the others do not. A group of type $C_2 \times C_4$ contains an element of order 4 but $C_2 \times C_2 \times C_2$ does not.

Now we look for non-abelian groups of order 8. Such a group must contain an element i of order 4 (else every element would square to 1 and so G would be abelian) and no element of order 8 (else G would be cyclic and hence abelian). Let $A = \langle i \rangle$. Since $|G : A| = 2$ it follows that $A \triangleleft G$. If $G - A$ contains j of order 2, then $i^j = i^{-1}$ else $i^j = i$ and G would be abelian. It follows that $G \simeq D_8$. It remains to consider what happens if every element $j \in G - A$ has order 4. Let $B = \langle j \rangle$, so $N = A \cap B$ must be central, and by result (c) it follows that $|N| = 2$.. The factor group G/N is not cyclic, else G would be be abelian by the solution to Exercise 2.13, p.53. Thus $G/N \simeq C_2 \times C_2$. Now $k = ij \notin A$ and so has order 4. Note that $G = A \cup Aj$ and that $A = \langle i \rangle$ is cyclic.

The structure of G will be determined once we can express ji as an element of
of $A \cup Aj$. Note that G contains a unique element of order 2 so $ijij = j^2$ so
$ji = i^3 j^2 j^3 = i^3 j = zij$ where $z = i^2 = j^2 = k^2$ is central. We elaborate for
a moment on why it is that the group structure of G is fully determined. We
have shown that every element of G is of the form $i^m j^n$ where $m \in \{0, 1, 2, 4\}$
and $n \in \{0, 1\}$. To multiply any two such expressions, you juxtapose them, and
then successively replace any occurrence of ji by $i^3 j$. For example

$$(ij)(i^2 j) = i(ji)ij = i(i^3 j)ij = jij = (ji)j = (i^3 j)j = i^3.$$

Once you know how to multiply elements, you determine the identity element
(in our case $i^0 j^0$) and inverses.

Thus there is at most one group missing from our list of isomorphism types
of groups of order 8. We have done enough work to show that if such a group
exists, we know how its elements multiply. However, at this stage there is no
reason to believe that this fifth type of group of order 8 actually exists. There
is the possibility that if we did some more reasoning, we might discover that
a contradiction flowed from the supposed existence of this group. To settle the
matter, we exhibit a concrete (i.e. a specific and reassuringly tangible) group
which has a multiplication along the lines that we have described.

To this end we consider the subgroup of S_8 generated by the following pair
of permutations:

$$I = (1, 2, 3, 4)(5, 6, 7, 8) \text{ and } J = (1, 5, 3, 7)(2, 8, 4, 6).$$

It follows that $K = IJ = (1, 8, 3, 6)(2, 7, 4, 5)$ and we put $Z = I^2 = J^2 = K^2 = (1, 3)(2, 4)(5, 7)(6, 8)$. A mechanical calculation shows that

$$Q = \langle I, J \rangle = \{ \text{id}, I, Z, IZ, J, K, JZ, KZ \}$$

is the fifth group which we sought. Now Q is always called the group of quaternions. It can also be expressed as a collection of non-singular 2×2 matrices
with entries which are complex numbers. This raises the terrifying thought that
we will have two meanings for the symbol i. It was an element of the group of
quaternions (just before the quaternions became permutations), and simultaneously it is a complex number which may crop up at any moment as a matrix
entry. So, forget the old usage of i for a moment, and recycle notation abusively
(but change the font to assuage the inevitable guilt). We put

$$\mathbf{i} = \begin{pmatrix} -i & 0 \\ 0 & i \end{pmatrix}, \mathbf{j} = \begin{pmatrix} 0 & 1 \\ -1 & 0 \end{pmatrix}, \mathbf{k} = \begin{pmatrix} 0 & -i \\ -i & 0 \end{pmatrix}$$

then the group of quaternions is isomorphic to $\{\pm I, \pm \mathbf{i}, \pm \mathbf{j}, \pm \mathbf{k}\}$. Now $-I$ is
playing the role of the permutation Z, and the other correspondences are the

obvious ones. William Rowan Hamilton thought of the quaternions (as a generalization of complex numbers) while walking in Dublin, and carved the key
equations defining the multiplication of $\mathbf{i}, \mathbf{j}$ and $\mathbf{k}$ on Brougham Bridge. The
devoted reader may wish to make a pilgrimage.

In terms of this representation as matrices, the group of quaternions consist
of 8 elements in a 4-dimensional $\mathbb{R}$-vector space. The matrices $I, \mathbf{i}, \mathbf{j}, \mathbf{k}$ are a
basis of this space. The real spans of $\{I, \mathbf{i}\}$, $\{I, \mathbf{j}\}$ and $\{I, \mathbf{k}\}$ are all copies of
the complex numbers with $\mathbf{i}, \mathbf{j}$ and $\mathbf{k}$ playing the role of a square root of -1,
and I playing the role of 1. Note that $\mathbf{ij} = \mathbf{k}$ etc, and indeed $\mathbf{ijk} = -I$. Viewed
as symbols rather than matrices, it is harmless to write 1 for I.

Hamilton's vision was that the quaternions would have endless uses in me-
chanics by letting the real span of 1 represent time, and the 3-dimensional space
spanned by $\mathbf{i}, \mathbf{j}$ and $\mathbf{k}$ represent *space* in the physical sense. In fact this scheme
was not a complete success, but this 4-dimensional space remains an structure
of considerable interest, especially to pure mathematicians.

Order 9: Since $9 = 3^2$ is the square of a prime number, any group of order 9
must be either a copy of C_9 or of $C_3 \times C_3$.

Order 10: Since $10 = 2 \times 5$ is twice an odd prime number, every group of order
10 must either be C_{10} or D_{10}. The latter group is not abelian, whereas C_{10} is
abelian.

Order11: Since 11 is a prime number, any group of order 11 is a copy of C_{11}.

Order 12: The structure theorem shows that there are two types of abelian
group of order 12; they are C_{12} and $C_2 \times C_6$. The obvious non-abelian groups
are the alternating group A_4, the dihedral group D_{12} and a semi-direct product
of C_3 by C_4. The final group can be realised as a subgroup of S_7 by putting
$a = (5, 6, 7)$ and $b = (1, 2, 3, 4)(6, 7)$. The cyclic group $A = \langle a \rangle$ is normal. These
three non-abelian groups are distinct because A_4 has no subgroup of order
6, whereas the other two groups each have cyclic subgroups of order 6. To
distinguish the remaining pair of groups, notice that the semi-direct product
has an element of order 4, whereas D_{12} does not. The alert reader may be
wondering about $S_3 \times C_2$, but as you should check, this is isomorphic to D_{12}.

Now we show that these three types of non-abelian group are the only ones
of order 12 which can arise. There must be a cyclic C subgroup of order 3 by
Cauchy's theorem. The action of G on the right cosets of C has kernel $\text{core}_G(C)$
by Remark 3.37. Unless $C \lhd G$, the group C must differ from at least one of its
conjugates, so $\text{core}_G(C)$ will be a proper subgroup of C. However C has order 3
so this forces $\text{core}_G(C) = 1$. The kernel of the action is trivial and so the induced
homomorphism from G to S_4 is a monomorphism (see Definition 3.23). This
image of this map has order 12 and is therefore of index 2 in S_4. Thus $G \simeq A_4$

by Proposition 3.24. Thus we may assume that $C \lhd G$. If there is an element of order 4 in G, then since G is non-abelian, conjugation by this element must invert elements of C, and we have the semidirect product. Finally if there are no elements of order 4, then all elements of G which are not in C have order 2 or 6. There are 3 Sylow 2-subgroups of G, else G would be abelian. If V_1, V_2 are distinct 3 Sylow 2-subgroups then since $4^2 > 12$ it follows that $V_1 \cap V_2 = \langle x \rangle$ is a subgroup of order 2. Since each V_i is abelian and $\langle V_1, V_2 \rangle = G$, it follows that x is central. Let $C = \langle c \rangle$ and suppose that $y \in V_1 - \langle x \rangle$. Now cx has order 6, y is an involution and $c^y = c^{-1}$. Thus $G \simeq D_{12}$.

The centralizer $D = C_G(C)$ has order 6, and so must be cyclic since it is not S_3. If $x \in G - D$ then D has order 2 and conjugation by x inverts every element of D, so $G \simeq D_{12}$.

Order 13: Since 13 is a prime number, any group of order 13 is a copy of C_{13}.

Order 14: Since $14 = 2 \times 7$ is twice an odd prime number, every group of order 14 must either be C_{14} or D_{14}. These groups are not isomorphic for reasons we have explained earlier.

Order 15: Let G be a group of order 15. There is $a \in G$ of order 3 and $b \in G$ of order 5 by Cauchy's theorem. Let $A = \langle a \rangle$ and $B = \langle b \rangle$. Observe that A, B are both normal in G by result (e), and are the only non-trivial Sylow subgroups of G. If $ab \in A \cup B$ then either $b \in A$ or $a \in B$ which is absurd. Now the order of ab can be neither $1, 3$ nor 5 and so $\langle ab \rangle = G$ and therefore $G \simeq C_{15}$.

A computer library of small groups is available, and the interested reader should try to obtain a copy of GAP or MAGMA. These are computer algebra systems which focus on group theory. See this book's web site for the links. There are remarkably many groups of 2-power order, although Mike Newman and Eamonn O'Brien have become past masters at enumerating them. There are approximately 5×10^{10} non-isomorphic groups of order $1024 = 2^{10}$. For the record the exact number is $49,487,365,422$. Besche, Eick and O'Brien constructed a data library for use with these computer algebra systems which contains all of the groups of order at most 2000 with the exception of those of order 1024: of the $423,164,062$ (pairwise non-isomorphic) groups stored there, the overwhelming majority have order 1536.

We now change tack a little. Instead of trying to understand all groups of a given order, we will make some tentative steps towards understanding the non-abelian finite simple groups. The phrase "scratching the surface" does not capture the modesty of the impression that we will make.

Beginning at the beginning, we show that there are no non-abelian finite simple groups G of order less than 60. Now, there are depressingly many natural numbers less than 60, so it makes sense to try to dispose of great chunks of numbers as non-starters.

Proposition 4.1

Groups of order p^n, pq (p and q distinct primes) cannot be non-abelian simple groups.

Proof

The trivial group is not simple by definition. If G is a group of order p^n with p prime and $n \geq 1$, then $Z(G) \neq 1$ by Proposition 3.38. However $Z(G) \trianglelefteq G$ so the simplicity of G forces $G = Z(G)$ to be abelian. If G has order pq then we may assume that $p > q$. Sylow's theorem ensures that G has a unique subgroup of order p.

□

Proposition 4.2

Suppose that the finite group G is such that $|G| = pr$ where p is a prime number and $r < p$, then G is not a non-abelian finite simple group.

Proof

Sylow's theorem forces $|\mathrm{Syl}_p(G)| = 1$ and simplicity forces $\mathrm{Syl}_p(G) = \{G\}$ and $r = 1$. Now G has prime order and so is abelian.

□

Proposition 4.3

There is no non-abelian finite simple group G of order less than 60.

Proof

Let G be the group under scrutiny. We may deploy Propositions 4.1 and 4.2. The positive integers less than 60 which are still candidate orders for non-abelian simple groups are as follows:

$$12, 18, 24, 30, 36, 40, 45, 48, 50, 54 \text{ and } 56.$$

Now the orders 18, 50 and 54 entail a Sylow subgroup of index 2 which must therefore be normal in G. Next we deploy Poincaré's argument [Proposition 3.35] on some of the other remaining candidate orders. In groups of order

12, 24 and 48 a Sylow 2-subgroup has index 3 but $12, 24$ and 48 do not divide $3! = 6$. In a group of order 36 a Sylow 3-subgroup has index 4 but 36 does not divide $4! = 24$. In a group of order 45 a Sylow 3-subgroup has index 5 but 45 does not divide $5! = 120$. The remaining candidate orders are 30, 40 and 56.

A simple group of order 30 would have to contain 6 Sylow 5-subgroups, and therefore 24 elements of order 5. It would also have to contain 10 Sylow 3-subgroups and so 20 elements of order 3. Now $24 + 20 > 30$ so this is absurd. In a group of order 40 the number of Sylow 5-subgroups must be 1 by Sylow's theorem, so it cannot be simple. In a simple group of order 56 there must be 8 Sylow 7-subgroups and so 48 elements of order 7. The remaining $56 - 48 = 8$ elements of the group must comprise the unique Sylow 2-subgroup, which must therefore be normal which is absurd.

$\square$

In fact there is a non-abelian group of order 60, and the second smallest non-abelian finite simple group has order 168.

Proposition 4.4

The alternating group A_5 is a simple group of order 60.

Proof

The group A_5 has index 2 in S_5 and so has order $5!/2 = 60$. There are 24 5-cycles in A_5, and therefore 6 Sylow 5-subgroups. If a normal subgroup contains just one 5-cycle, it must contain all its powers and their conjugates, and therefore all Sylow 5-subgroups of A_5, and hence all 24 5-cycles. Therefore the only multiple of 5 which is a candidate order for a proper normal subgroup of A_5 is 30. However, a subgroup of order 30 would have to contain an element of order 3, and therefore all 20 of them (by normality). Indeed, this argument eliminates the remaining multiples of 3 as candidates for orders of proper normal subgroups. Thus the only possible orders for a proper normal subgroup of A_5 are 2 and 4. However, the Sylow 2-subgroups of A_5 are the conjugates of $V = \{e, (12)(34), (13)(24), (14)(23)\}$. However, V stabilizes 5 and V^g stabilizes $(5)g$ for every $g \in A_5$. Now A_5 acts transitively on $\{1, 2, 3, 4, 5\}$ so the intersection of the conjugates of V is trivial. However, a normal subgroup of order 2 or 4 would have to be contained in one and therefore all of these conjugates by Sylow's theorem. Thus A_5 is simple.

$\square$

Before reading further, now that you have read some of the techniques which can be deployed in this elimination game, you are strongly urged to attempt the proof of the next result. If you don't have the stomach for the whole thing, pick some numbers in the range at random. You may find 63, 90, 112, 120 and 144 particularly entertaining.

Proposition 4.5

There is no non-abelian finite simple group G with $60 < |G| < 168$.

Proof

Once again we may deploy Propositions 4.1 and 4.2 to eliminate some possibilities. The remaining candidate orders are:

$$63, 70, 72, 75, 80, 84, 90, 96, 98, 100, 105, 108, 112, 120, 126,$$

$$132, 135, 140, 144, 147, 150, 154, 160, 162 \text{ and } 165.$$

Now (normal) Sylow subgroups of index 2 arise when the order is 98 and 162. Poincaré's theorem will come into play for the order 75 (since 75 does not divide $3! = 6$), 80 (because 80 does not divide $5!=120$), 96 (since 96 does not divide $3! = 6$), 100 (since 100 does not divide $4! = 24$), 108 (because 108 does not divide $4! = 24$), 135 (since 135 does not divide $5! = 60$), 147 (because 147 does not divide $3! = 6$), 150 (since 150 does not divide $6! = 720$) and 160 (because 160 does not divide $5! = 120$). The surviving candidate orders are

$$63, 70, 72, 84, 90, 105, 112, 120, 126, 132, 140, 144, 154, 165.$$

In G a simple group of order 63 there must be 7 Sylow 3-subgroups. Choose P, Q distinct Sylow 3-subgroups and put $T = P \cap Q$. Now $|PQ| \le 63$ so $|P||Q|/|T| \le 63$. Thus $|T| = 3$. Now T is a proper subgroup of each of the finite 3-groups P and Q, and so must be properly contained in its normalizer in each of them. However, T is a maximal subgroup of each of P and Q, so $P, Q \le N_G(T)$ by Proposition 3.40. Thus $\langle P, Q \rangle \le N_G(T)$, but P is a maximal subgroup of G and Q is not a subgroup of P so $\langle P, Q \rangle = G$. Thus $T \lhd G$ and we have a contradiction.

That argument is too good to waste. By replacing 3 by a prime number p and 7 by the prime number $q \ne p$, it shows that there is no simple group of order $p^2 q$ whenever $p^2 > q$. On the other hand, if $p^2 < q$ then $q > \sqrt{|G|}$, so G is not simple. Thus no simple group can have order $p^2 q$.

Any group of order $70 = 5 \cdot 14$ must have a normal subgroup of order 5 by Sylow's theorem.

Now suppose that G is a simple group of order $72 = 2^3 3^2$. By Sylow's theorem there must be exactly 4 conjugate subgroups of order 9. Let P be one of them, so $|G : N_G(P)| = 4$. Now Poincaré's argument forces 72 to divide $4! = 24$ and this disposes of 72.

Next suppose that G is a group of order $84 = 2^2 \cdot 3 \cdot 7$. Sylow's theorem tells us that there is a unique Sylow 7-subgroup which must be normal.

Now suppose that G is a simple group of order $90 = 2 \cdot 3^2 \cdot 5$. There must be 6 Sylow 5-subgroups, and 10 Sylow 3-subgroups. There are 24 elements of order 5. If each pair of distinct Sylow 3-subgroups intersects in the trivial group then there would be 80 elements of order 3 or 9. Now $24 + 80 > 90$ so there are $P, Q \in \mathrm{Syl}_3(G)$ with $T = P \cap Q$ of order 3. Let $S = N_G(T)$, so $P, Q < S$. Thus $|S| = 18, 45$ or 90. Now T is not normal in G by simplicity, so $|S| \neq 90$. If $|S| = 45$, then $|G : S| = 2$, so $S \lhd G$ which is absurd so $|S| = 18$. Now Poincaré's theorem applies and we deduce that 90 divides $5! = 120$ which is absurd.

Next suppose that G is a simple group of order $105 = 3 \cdot 5 \cdot 7$. By Sylow's theorem there must 15 subgroups of order 7 and so 90 elements of order 7. Sylow's theorem also tells us that there are 21 subgroups of order 5 and therefore 84 elements of order 5 in G. Now $90 + 84 > 105$ so this is absurd.

Suppose that G is a simple group of order $112 = 2^4 \cdot 7$. There must be 7 Sylow 2-subgroups and 8 Sylow 7-subgroups. Let P, Q be distinct Sylow 2-subgroups with intersection T of maximal order. Since $16^2/|T| \leq 112$ it follows that $|T| \geq 256/112$ so $|T| = 4$ or 8. If T has order 8, then both P and Q normalize T so T is normal in $\langle P, Q \rangle = G$. This is absurd so $|T| = 4$. Let $S = N_G(T)$. Suppose that $P \leq S$, then because $S \cap Q$ is not contained in the maximal subgroup P, $T \lhd \langle P, S \cap Q \rangle = G$ which is absurd. Let $\widehat{P} = P \cap S$ and $\widehat{Q} = Q \cap S$. These groups $\widehat{P}, \widehat{Q}$ both have order 8 and normalize $T = \widehat{P} \cap \widehat{Q}$. Now $R = \langle \widehat{P}, \widehat{Q} \rangle$ normalizes T, and so cannot be G and has order strictly larger than 8. Thus $|R| = 16$ or 56. Now T is not normal in P but T is normal in R so $P \neq R$. Also $\widehat{P}$ has order 8 and is a subgroup of both P and R. If $P, R \in \mathrm{Syl}_2(G)$, then this violates the maximality of the order of the intersection of $P \cap Q$. Finally we conclude that $|R| = 56$ so $|G : R| = 2$ and therefore $R \lhd G$ which is absurd.

Suppose (for contradiction) that G is a simple group of order $120 = 2^3 \cdot 3 \cdot 5$. There must be 6 Sylow-5 subgroups, so if L is the normalizer of a subgroup of order 5, then $|G : L| = 6$. Now the natural action of G on $L \backslash G$ yields a monomorphism from G into S_6. Let the image of this map be $\widehat{G} \simeq G$. Now $A_6 \cap \widehat{G} \unlhd \widehat{G}$ and $|\widehat{G} : A_6 \cap \widehat{G}| = |\widehat{G} A_6 : A_6| \leq 2$. The simplicity of $\widehat{G}$ forces $\widehat{G} < A_6$. Let H be the stabilizer of 6 in A_6 in its natural action on $\{1, 2, 3, 4, 5, 6\}$. Now A_6 acts transitively and primitively on this set by the answer to Exercise 3.16, p. 122 combined with Proposition 3.52. Proposition 3.54 applies so H is a maximal subgroup of A_6. Now

$$|\widehat{G}H| = 120 \times 60/|H \cap \widehat{G}| \leq 360$$

so $|H \cap \widehat{G}| \geq 20$. However, $H \simeq A_5$ cannot have a proper subgroup of index 3 or less since it is simple and Poincaré's argument applies. Thus $H \cap \widehat{G} = H$ so $H < \widehat{G} < A_6$. This violates the maximality of H in A_6 and we are done.

Suppose that G is a group of order $126 = 2 \cdot 3^2 \cdot 7$. By Sylow's theorem there must be a unique subgroup of order 7.

Now suppose that G is a simple group of order $132 = 2^2 \times 3 \times 11$. There must be 12 Sylow 11-subgroups and so 120 elements of order 11. The group is not large enough to allow 22 Sylow 3-subgroups and so there are 4 of them. There must be 8 elements of order 3. Now $120 + 8 = 128$ and so the remaining 4 elements must comprise the unique Sylow 2-subgroup which must be normal. This contradicts simplicity.

Suppose that G is a group of order $140 = 2^2 \cdot 5 \cdot 7$. By Sylow's theorem there must be a unique subgroup of order 7.

Suppose that G is a simple group of order $144 = 2^4 \cdot 3^2$. Consider the Sylow 3-subgroups. By Sylow's theorem there must be 4 or 16 of them. However, if there were just 4, then the normalizer L of a Sylow 3-subgroup would have index 4 in G, and Poincaré's argument shows that 144 divides $4! = 24$ which is absurd. Therefore there are 16 Sylow 3-subgroups. If each pair of distinct Sylow 3-subgroups intersect in the trivial group, then G has 128 elements whose order is 3 or 9. The remaining 16 elements must comprise the unique Sylow 2-subgroup which violates simplicity. Therefore there are distinct Sylow 3-subgroups P, Q such that $P \cap Q = T$ has order 3. Now $P, Q < N_G(T)$, so $|N_G(T)| > 9$ and 9 divides $|N_G(T)|$. Now $|N_G(T)| \neq 18$ since any group of order 18 has a unique Sylow 3-subgroup.

Thus $|N_G(T)| = 36, 72$ or 144. The last option does not arise since it would entail $T \triangleleft G$. However, Poincaré's argument will not allow $|N_G(T)| = 36$ since 144 does not divide $4! = 24$, and $|N_G(T)| = 72$ is impossible since then $|G : N_G(T)| = 2$ so $N_G(T) \triangleleft G$.

Suppose that G is a group of order $154 = 2 \cdot 7 \cdot 11$. By Sylow's theorem there must be a unique subgroup of order 11.

Next suppose that G is a group of order $165 = 3 \cdot 5 \cdot 11$. By Sylow's theorem there must be a unique subgroup of order 11.

$\square$

There is a highly relevant result which would be very useful if we could deploy it in this quest, but unfortunately there is no known proof which is sufficiently elementary for inclusion here. This is Burnside's $p^\alpha q^\beta$ theorem.

Theorem 4.6 (Burnside)

Suppose that G is a finite group of order $p^\alpha q^\beta$, where p, q are distinct primes and α, β are non-negative integers. It follows that G is not a non-abelian simple group.

The fastest known proof involves something quite magical called *group representation theory*. This involves studying homomorphisms from finite groups G into matrix groups $\mathrm{GL}_n(\mathbb{C})$. Put like that, representation theory may sound rather grim, but it isn't. If you have been lucky or sensible enough to learn the theory of Fourier series (or better yet, proper Fourier analysis), you may have seen gems such as

$$1 + \frac{1}{2^2} + \frac{1}{3^2} + \cdots = \frac{\pi^2}{6}.$$

The theory of Fourier series consists of the study of moderately well-behaved periodic functions on $\mathbb{R}$. However, there is another way to say this; it consists of the study of moderately well-behaved functions on the circle (wrapping $\mathbb{R}$ around). Now all circles are similar, and happily one of them has an important group structure. This is the unit circle in the Argand diagram, corresponding to complex numbers of unit modulus. The group law is complex multiplication

$$e^{i\theta} \times e^{i\psi} = e^{i(\theta + \psi)}.$$

Of course this group is infinite, which is a shame from our current and very temporary point of view. However, topologically speaking (if you happen to speak topology) a circle is a compact set, and compact means *morally finite*. The theory of Fourier series does for the group of the unit circle what representation theory does for finite groups. If you have studied Fourier series you will know that many delicate analytic questions arise. The representation theory of finite groups does not really involve these analytic matters (because a finite group does not have an interesting topology). However, the pretty algebra survives intact.

The representation theory of finite groups was developed as the 19th century turned into the 20th century by Burnside and Frobenius, and has proved a powerful tool in the attempts to understand finite groups.

In 1962 an remarkable theorem (with what seemed at the time a very long proof) set off a group theoretic earthquake.

Theorem 4.7 (Feit-Thompson)

Let G be a non-abelian finite simple group, then $|G|$ is an even number.

Proof

Omitted.

□

A slightly earlier result is extremely relevant. Recall that the word *involution* is just snazzy terminology for an element of order 2.

Theorem 4.8 (Brauer-Fowler)

There is a function $f : \mathbb{N} \to \mathbb{N}$ with the following property. Let G be a non-abelian finite simple group containing an involution t. It follows that $|G| \leq f(|C_G(t)|)$.

Proof

Omitted.

□

Thus the order of a finite simple group is bounded above by a function of the order of the centralizer of an involution. Now the Feit–Thompson theorem combines with Cauchy's theorem to ensure that every non-abelian finite simple group contains an involution. A route to the classification of finite simple groups thus opened up, and the assault was made by trying to understand which non-abelian finite simple groups could have a centralizer of an involution of a certain type.

Let us return to the smallest non-abelian finite simple group. In fact up to isomorphism, A_5 is the only simple group of order 60, so there is no pretender its title. In principle this could be verified by inspection of all possible 60×60 multiplication tables, but for some reason to do with $60^{3,600}$ we shy away from this direct approach. Of course A_5 exists in many different guises. For example, it arises as the group of rigid symmetries (symmetries you can accomplish using rigid motions) of two of Plato's regular solids, the icosahedron and the dodecahedron. For this reason A_5 sometimes masquerades under the soubriquets *the icosahedral group* and *the dodecahedral group*.

We establish this uniqueness result by a sequence of lemmas.

Lemma 4.9

Suppose that G is a simple group of order 60. It follows that G has a subgroup H of order 12.

Proof

Observe that $|G| = 60 = 2^2 \cdot 3 \cdot 5$. By Sylow's theorem the number of Sylow 2-subgroups must be $3, 5$ or 15, but there cannot be 3 Sylow 2-subgroups by Poincaré's argument applied to the normalizer of a Sylow 2-subgroup. If there are 5 Sylow 2-subgroups, then the normalizer of any one of them will have index 5 and so be of order 12 as required. There remains the possibility that there are 15 Sylow 2-subgroups.

There are 6 Sylow 5-subgroups and so 24 elements of order 5. It follows that if there are 15 Sylow 2-subgroups then at least two of them must intersect in a group T of size 2 by counting (if all pairs of distinct Sylow 2-subgroups intersect trivially there are 45 elements of order 2 or 4; note that $24 + 45 > 60$). Therefore the normalizer in G of T must contain two different subgroups of size 4 and so be of order 12 or 20. The latter is impossible since a subgroup of index 3 in G would, thanks to Poincaré's argument, force 60 to divide $3! = 6$, so $|N_G(T)| = 12$.

$\square$

Lemma 4.10

Suppose that G is a simple group of order 60. It follows that G is isomorphic to a subgroup $\widehat{G}$ of S_5.

Proof

By Lemma 4.9 G has a subgroup H of order 12 and index 5. Let $\Omega = H\backslash G = \{Hx \mid x \in G\}$ so $|\Omega| = 5$. There is a natural action of G on Ω by right multiplication. If $A \in \Omega$ and $g \in G$, then $A * g = Ag$.

Any group action induces a group homomorphism, and ours is no exception. Let the homomorphism be $\theta : G \to \operatorname{Sym} \Omega \simeq S_5$. Therefore $\operatorname{Ker} \theta \leq H$ and the simplicity of G forces $\operatorname{Ker} \theta = 1$. Thus θ is a monomorphism, and via the isomorphism $\operatorname{Sym} \Omega \simeq S_5$ we see that G is isomorphic to $\widehat{G} < S_5$.

$\square$

We have seen the following result before as Proposition 3.24 but it merits repetition.

Lemma 4.11

Let $n \geq 2$ be a natural number. Suppose that $A < S_n$ and $|S_n : A| = 2$, then $A = A_n$.

Proof

Since A has index 2 in S_n we have $A \triangleleft S_n$. If $\tau \in S_n$ is any transposition and $\tau \in A$, then all transpositions are in A because all transpositions are conjugate in S_n. However, the transpositions generate S_n by Proposition 3.14 so this would force $A = S_n$ which is not the case. Therefore every transposition τ has the property that $A\tau \neq A$. The group G/A is cyclic of order 2. If $\sigma \in A_n$, then σ is the product of an even number of transpositions, so $A\sigma$ is an even power of the generator of G/A, and so $A\sigma = A$ and therefore $\sigma \in A$. This is true for any $\sigma \in A_n$, so $A_n \leq A$. However, both these groups have order $n!/2$, so $A = A_n$.

$\square$

Proposition 4.12

Let G be a simple group of order 60, then $G \simeq A_5$.

Proof

By Lemma 4.10 there is $A < S_5$ with $A \simeq G$. Now $|S_5 : A| = 2$ so by Lemma 4.11, $A = A_5$ and therefore $G \simeq A_5$.

$\square$

Thus we now know that up to isomorphism there is a unique non-abelian simple group of order 60, and we have not had to contemplate $60^{3,600}$ multiplication tables after all. Now we demonstrate the simplicity of A_n whenever n is large enough. In the argument which follows we use the notation $\mathrm{Stab}_{A_n}(\omega)$ for the stabilizer of $\omega \in \{1, 2, \ldots, n\}$ in A_n. This is to avoid terminological overcrowding. After all, A_n looks worryingly like the stabilizer (isotropy group) in a group called A of the point $n \in \{1, 2, \ldots, n\}$.

Proposition 4.13

The group A_n is simple whenever $n \geq 5$.

Proof

We proceed by induction on $n \geq 6$ since A_5 is known to be simple.

Suppose (for contradiction) that A_n has a non-trivial proper normal subgroup M. Now A_n is a permutation group on $\Omega = \{1, 2, \ldots, n\}$, and acts at least 4-transitively on this set by Exercise 3.16, p. 122. The stabilizer of each $\omega \in \Omega$ is a copy of A_{n-1}. Now Propositions 3.52 and 3.54 apply so $\mathrm{Stab}_{A_n}(\omega)$ is a maximal subgroup since A_n acts 2-transitively on Ω. These groups $\text{Stab}_{A_n}(\omega)$ are conjugate. If any one of them were contained in M, then they all would be (by conjugacy), and by maximality $M = A_n$ which is absurd. Now each $\text{Stab}_{A_n}(\omega)$ is simple (by induction), so each $M \cap \text{Stab}_{A_n}(\omega) = 1$. Maximality now ensures that each Thus $M\text{Stab}_{A_n}(\omega) = A_n$. Now $\text{Stab}_{A_n}(\omega) \simeq A_{n-1}$. Therefore

$$n!/2 = |A_n| = |M\text{Stab}_{A_n}(\omega)| = |M|(n-1)!/2$$

so $|M| = n$.

There are a variety of ways to clinch the argument from here, but there is still a little work to do (for a reason that we will explain at the end of the proof). Suppose that $\tau \in S_n$ Conjugation by τ induces an automorphism of A_n, so $M^\tau \lhd S_n$. Suppose (for contradiction) that $M^\tau \neq M$. Now MM^τ will be a normal subgroup of A_n which properly contains M, so $|MM^\tau| > n$. We have shown that any non-trivial proper normal subgroup of A_n has order n, so $MM^\tau = A_n$. Now $|MM^\tau| \leq n^2$ by Corollary 2.31, so $|A_n| = n!/2 \leq n^2$ which is false by induction whenever $n \geq 5$. Thus $M^\tau = M$ for all $\tau \in S_n$. Therefore $M \lhd S_n$. Thus M of size n is a union of conjugacy classes of S_n. and by Cauchy's theorem we may choose $m \in M$ of prime order p. The cycle shape of m involves no 1-cycle, since M intersects each $\text{Stab}_{A_n}(\omega)$ in the trivial group. Thus m is a product of r disjoint cycles of length p where $rp = n$. If $p > 2$, then there are at least distinct $(n-1)(n-2)$ cycles of length p of the form $(1, a, b, \ldots)$ since we may choose a in $n - 1$ ways and then b in $n - 2$ ways. Now all elements of the same cycle shape are conjugate in S_n and so M contains at least $(n-1)(n-2) + 1$ elements (including the identity). Now $(n-1)(n-2) + 1 > n$ for $n \geq 3$ by an induction argument, so this is absurd.

We conclude that $p = 2$. We now seek a lower bound for the number of elements in S_n with cycle shape consisting of r disjoint 2-cycles where $n = 2r$. If $n \geq 6$, then there are at least $(n-3)(n-4)(n-5)$ elements of S_n of shape $(1, a)(2, b)(3, c) \cdots$. Now $(n-3)(n-4)(n-5)+1 > n$ for $n \geq 6$ by an induction argument. We finally have the contradiction which clinches the proof.

$\square$

Note that A_3 is cyclic of order 3, and is therefore simple. However, A_4 is not simple, and in fact has the normal subgroup

$$\{ \text{id}, (1,2)(3,4), (1,3)(2,4), (1,4)(2,3) \}.$$

This is a normal subgroup of order 4 of the group A_n when $n = 4$. Therefore the argument to clinch the proof must use the hypothesis that the degree under scrutiny is at least 6 (A_5 is the base case) in a crucial way. Hence the combinatorial fiddling at the end of the proof of Proposition 4.13.

Now is the time to lay the ghost of 120. We had to work rather hard to show that there was no simple group of order 120. Now we know that A_6 is a simple group, Poincaré's argument tells us that it can have no subgroup of index 3. Please look back to the argument given in Proposition 4.5 which eliminated 120 as a candidate order of a finite simple group, and work out how to shorten the proof.

A wide variety of results about finite groups can be obtained with the tool-box that we have assembled. In the days before the classification of finite simple groups, it was good drill to use these results to show that there are no finite simple groups of order n where n is a specified natural number for which the result is true. A variation is to select m which is the order of a finite simple group, and use the tool-box to show that up to isomorphism there is just one finite simple group of order m in the event that this result is true for the particular natural number m. These activities seem a little pointless in view of the fact that we have a list of the finite simple groups. Nonetheless, these games are fun, and we give a couple of examples of these proofs to illustrate the scope for ingenuity that such questions provide, and for sentimental reasons.

So, if you have developed a taste for these activities, the next example demonstrates what can be done using sustained and determined application of the results in our tool-box. However, you should feel free to skip it if you have had enough of Sylow arguments (a strong stomach is required).

Proposition 4.14

If G, H are both simple groups of order 168, then $G \simeq H$.

Proof

Let G be a simple group of order $168 = 2^3 \cdot 3 \cdot 7$. Now 168 is not a prime number, so G is non-abelian. Thanks to the Sylow Theorems we may conclude that $|\mathrm{Syl}_2(G)| = 3, 7$ or 21, $|\mathrm{Syl}_3(G)| = 7$ or 28, and finally $|\mathrm{Syl}_7(G)| = 8$. In fact we can immediately eliminate the possibility that $|\mathrm{Syl}_2(G)| = 3$ by Poincaré's argument.

We will now show that $|\mathrm{Syl}_3(G)| = 28$. To this end, we assume (for contradiction) that $|\mathrm{Syl}_3(G)| = 7$. Choose $P \in \mathrm{Syl}_7(G)$ so $|G : N_G(P)| = 8$. Thus $|N_G(P)| = 21$. Apply Sylow's Theorem to the Sylow 3-subgroups of $N_G(P)$. Thus $N_G(P)$ contains either 1 or 7 subgroups of order 3. We will show that each possibility leads to an absurdity.

If $N_G(P)$ contains a unique Sylow 3-subgroup Q, it follows that $N_G(P) \leq N_G(Q)$. Now $|G : N_G(P)| = 8$ but $|G : N_G(Q)|$ is 7 or 28, neither of which is a divisor of 8. This is absurd.

Now we examine the possibility that $N_G(P)$ contains 7 Sylow 3-subgroups. Thus all Sylow 3-subgroups of G are contained in $N_G(P)$, and indeed $N_G(P)$ is generated by the set of all Sylow 3-subgroups of G. It follows that $N_G(P) \triangleleft G$, but G is simple. This too is absurd. We deduce that $|\mathrm{Syl}_3(G)| = 28$.

Let us summarize what we know of the Sylow structure of a simple group G of order 168.

Prime	$|X|$ where $X \in \mathrm{Syl}_p(G)$	$|\mathrm{Syl}_p(G)|$ $(= |G : N_G(X)|)$	$|N_G(X)|$
2	8	7 or 21	24 or 8
3	3	28	6
7	7	8	21

We will next show that $|\mathrm{Syl}_2(G)| = 21$, so we assume (for contradiction) that $|\mathrm{Syl}_2(G)| = 7$. Choose $Q \in \mathrm{Syl}_2(G)$ so $|N_G(Q)| = 24$. Now the number of distinct subgroups of order 3 in $N_G(Q)$ must be 1 or 4. However, it cannot be 1, else the normalizer in G of a Sylow 3-subgroup would have order at least 24, whereas we know such an normalizer must have order 6. Thus there are 4 distinct subgroups of order 3 in $N_G(Q)$. Note that by conjugacy of the Sylow 3-subgroups, every subgroup of G of order 3 must normalize a Sylow 2-subgroup. There are 7 distinct conjugates of Q in G, and so at most 7 conjugates of $N_G(Q)$, each of which contains 4 subgroups of order 3. However, $|\mathrm{Syl}_3(G)| = 28$ so there are indeed 7 distinct conjugates of $N_G(Q)$ no distinct pair of which contain a common element of order 3. It follows that if $Q_1, Q_2 \in \mathrm{Syl}_2(G)$ are distinct, then $N_G(Q_1)$ and $N_G(Q_2)$ are distinct. Now $R = N_G(Q_1) \cap N_G(Q_2)$ cannot contain an element of order 3, so $|R|$ is a power of 2.

Now

$$168 = |G| \geq |N_G(Q_1)N_G(Q_2)| = \frac{|N_G(Q_1)||N_G(Q_2)|}{|R|} = \frac{24^2}{|R|}.$$

Thus $|R| \geq 24^2/168 = 3\frac{3}{7}$ so R is a group of order 4 or 8. However, it cannot be that $|R| = 8$ for then $Q_1 = R = Q_2$, so $|R| = 4$ and R is a normal subgroup of both Q_1 and Q_2, since it has index 2 in each of them. Now $N_G(R)$ contains Q_1 and Q_2, and so 8 divides $|N_G(R)|$ which in turn divides 168, and moreover, $|N_G(R)| > 8$. By Poincaré's argument $|N_G(R)| \leq 24 = 168/7$ because the order of G must divide $|G : N_G(R)|!$ and 7 divides $|G|$. The only possibility is that $|N_G(R)| = 24$ and $N_G(R)$ contains at least 2, and therefore exactly 3, subgroups of order 8.

Recall that $N_G(Q)$ also has order 24, but has a unique subgroup of order 8, unlike $N_G(R)$ which contains 3 subgroups of order 8. By replacing Q by a suitable conjugate, we may assume that the distinct groups $N_G(Q)$ and $N_G(R)$ share a common subgroup of order 3. Now Q and any Sylow 2-subgroup of $N_G(R)$ meet in a group of order divisible by 4, so $|N_G(Q) \cap N_G(R)|$ divides 24, is not 24, and is divisible by 12. Put $H = N_G(Q) \cap N_G(R)$ so H has order 12, and so has index 2 in each of $N_G(R)$ and $N_G(Q)$. We conclude that $H \lhd \langle N_G(R), N_G(Q) \rangle$. However, the index of $\langle N_G(R), N_G(Q) \rangle$ in the simple group G is less than 7. We conclude that $\langle N_G(R), N_G(Q) \rangle = G$ by Poincaré's argument. Now $1 \neq H \lhd G$ which is absurd. At last we have eliminated the possibility that $|\mathrm{Syl}_2(G)| = 7$.

The state of play is now as follows:

prime	$|P|$	$|\mathrm{Syl}_p(G)|$	$|N_G(P)|$
2	8	21	8
3	3	28	6
7	7	8	28

We now examine how distinct elements of $\mathrm{Syl}_2(G)$ intersect; our first ambition is to show that there are $S_1, S_2 \in \mathrm{Syl}_2(G)$ with $|S_1 \cap S_2| = 4$. We can eliminate the possibility that every pair of distinct elements of $\mathrm{Syl}_2(G)$ intersect trivially, for in that event, there would be $21 \times 7 = 147$ non-identity elements of G having 2-power order, as well as $2 \times 28 = 56$ elements of order 3 in G. The group G has only $168 < 203 = 147 + 56$ elements, so this is absurd. Thus there are $T_1, T_2 \in \mathrm{Syl}_2(G)$ with $|T_1 \cap T_2| \in \{2, 4\}$. Let $U = T_1 \cap T_2$. If U has order 4 we are done, so we may assume that $|U| = 2$. Let $M_i = N_{T_i}(U)$ for $i = 1, 2$. By Proposition 3.40 each M_i has order 4 or 8. Let $V = \langle M_1, M_2 \rangle$ so $U \lhd V$. The group G is simple so $V \neq G$. Now 4 divides $|V|$ which in turn divides 24, and moreover $|V| > 4$. It follows that $|V| = 8, 12$ or 24.

(a) Suppose that $|V| = 8$. It follows that $V \cap T_1$ has order 4 and we are done.

(b) Suppose that $|V| = 12$. Now V has either 1 or 4 subgroups of order 3. The former is impossible by considering the size of the normalizer of such a subgroup, so V has 4 subgroups of order 3, and hence contains 8 elements of order 3. The remaining 4 elements must comprise the unique Sylow 2-subgroup of V. This forces $T_1 \cap V = T_2 \cap V$ which is absurd. Thus $|V| \neq 12$.

(c) Suppose that $|V| = 24$. If both $T_1, T_2 \leq V$, then

$$24 = |V| \geq |T_1 T_2| = |T_1||T_2|/|T_1 \cap T_2| = 64/2 = 32$$

which is absurd. We know that $M_1, M_2 \leq V$ so without loss of generality we may assume that $|T_1 \cap V| = 4$. By Sylow's theorem there is $S \in \mathrm{Syl}_2(V)$ with $T_1 \cap V \leq S$. Now $|S| = 8$ so $S \in \mathrm{Syl}_2(G)$ and $M_1 \leq T_1 \cap S \leq T_1 \cap V$. Now $4 \leq |M_1| \leq |T_1 \cap S| \leq |T_1 \cap V| = 4$. Thus $T_1 \cap S$ has order 4 and there is cause for minor celebration.

We now know that there is a pair S_1, S_2 of Sylow 2-subgroups of G which intersect in a subgroup of order 4. Since the intersection cannot be normal in G, it follows that $Y = \langle S_1, S_2 \rangle$ has order 24, where we have $|\mathrm{Syl}_2(Y)| = 3$ and $|\mathrm{Syl}_3(Y)| = 4$, and Y is self-normalizing. There are exactly 7 conjugates of Y in G, and since there are 28 Sylow 3-subgroups of G, it follows that the intersection of distinct conjugates of Y must be a 2-group. Moreover, there are 21 Sylow 2-subgroups of G, so each occurs in exactly one conjugate of Y. Any pair of distinct conjugates Y_1, Y_2 of Y will intersect in a subgroup W_{12} of order 4, and W_{12} will be contained in a unique Sylow 2-subgroup of each of Y_1 and Y_2. These 2-groups generate a group Z of order 24 having three Sylow P subgroups. Two of these we know about already. The third must be in a new conjugate of Y, say Y_3. Now Y_1, Y_2 and Y_3 are the only conjugates of Y sharing a Sylow 2-subgroup with Z. Notice that Z has 7 conjugates.

We now construct a "geometry" (a purely combinatorial matter). This involves specifying a set P of "points", and a collection L of "lines" and rules concerning which points and lines are incident.

The points of the geometry will be the conjugates of Y. The lines of the geometry will be the conjugates of Z. A point and line are incident (the point is on the line) if and only if the groups share a Sylow 2-subgroup. Every point is incident to three lines, and every line is incident to three points. See Figure 4.1.

The group G acts transitively by conjugation on the vertices of this geometry, preserving the incidence structure. However, a careful inspection yields that the full automorphism group of the geometry has order at most 168, and so is a copy of G. Thus any two simple groups of order 168 are isomorphic.

$\square$

Both $\mathrm{PSL}(2,7)$ and $\mathrm{GL}(3,2)$ are simple groups of order 168. If these names seem confusing, read on and we will explain. We have already discussed general linear groups in Chapter 1, so $\mathrm{GL}(3,2)$ may ring a bell. It is the multiplicative group of 3×3 invertible matrices with entries drawn from a field of size 2. A field of size 2 must be a copy of the integers modulo 2. It consists of two quantities 0 and 1 and the addition and multiplication is driven by $1 + 1 = 0$. You can think of 0 is been "even" and 1 as being "odd", and the algebra being that of odd and even numbers: odd + odd = even and so on. If finite fields are a lacuna in your education, we provide Appendix A.

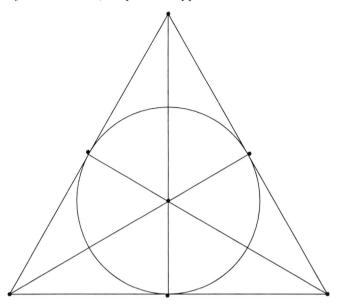

Figure 4.1 Seven points on seven lines

The group $\mathrm{PSL}(2,7)$ is a little more tricky to describe. You start with $\mathrm{GL}(2,7)$, the multiplicative group of invertible 2×2 matrices with entries drawn from a field with 7 elements. the determinant map yields an epimorphism from $\mathrm{GL}(2,7)$ to the cyclic group of order 6 (the multiplicative group of the field). Thus $\mathrm{GL}(2,7)$ is not simple. The kernel of the determinant homomorphism is $\mathrm{SL}(2,7)$, the special linear group of 2×2 matrices with entries in our field. Now $\mathrm{SL}(2,7)$ is not simple because it has a non-trivial centre Z which (you can verify) consists of $\pm I$ where I is the 2×2 identity matrix with entries in this field. The group $\mathrm{PSL}(2,7)$ is defined to be the factor group $\mathrm{SL}(2,7)/Z$. There is an essentially unique finite field of each prime power order (other than 1). Thus $\mathrm{PSL}(2,7)$ has cousins $\mathrm{PSL}(n,p^m)$ defined in an entirely analogous way. These are not quite groups of matrices as we have described them here, but they are factor groups of groups of matrices by easily understood central subgroups of scalars. Subject to some small exceptions ($\mathrm{PSL}(2,m)$ for $m = 2,3$) these groups are all non-abelian finite simple groups.

The classification of finite simple groups runs along the following lines (very crudely). There are the cyclic groups of prime order, there are the alternating groups A_n for $n \geq 5$. There are lots of families of groups which arise from studying fairly natural groups of matrices with entries in finite fields. Apart from these examples (which we have not listed in detail) there are 26 groups. These are called the sporadic finite simple groups. Five of them were discovered very early by Mathieu in 1861 and 1873, and the rest were found in the last century between 1964 and 1981.

EXERCISES

Let k be a finite field of order q. Up to isomorphism there is a unique finite field of each order q where $q \neq 1$ is a prime power. There are no other finite fields (see Appendix A).

4.1 Count the n by n matrices with entries in k.

4.2 An n by n matrix is in $\mathrm{GL}(n, q)$ precisely when its rows are linearly independent as elements of k^n. Derive a formula for $|\mathrm{GL}(n, q)|$ and verify that $|\mathrm{GL}(3, 2)| = 168$.

4.3 An n by n matrix is in $\mathrm{SL}(n, q)$ precisely when it is in the kernel of the determinant epimorphism det $: \mathrm{GL}(n, q) \to k^*$ where k^* is the multiplicative group of non-zero elements of k. Derive a formula for $|\mathrm{SL}(n, q)|$.

4.4 Let Z denote the group of scalar matrices (scalar multiples of the identity matrix) in $\mathrm{SL}(n, q)$. Suppose that the polynomial $X^n - 1$ has t roots in k. Define $\mathrm{PSL}(n, q)$ to be $\mathrm{SL}(n, q)/Z$. Derive a formula for $|\mathrm{PSL}(n, q)|$ and verify that $\mathrm{PSL}(2, 7)$ has order 168.

$\mathrm{GL}(3, 2)$

Thanks to the solution to Exercise 4.2 above, we now know that $|\mathrm{GL}(3, 2)| = 168$. Suppose (for contradiction) that $1 \neq N \lhd G$. Among all such normal subgroups N, we choose one of maximal order. Now G/N must be a simple group, else by the Correspondence Principle we could insinuate a normal subgroup properly between N and G. Now the only non-abelian simple group of order less than 168 is A_5, a group of order 60. Now 60 does not divide 168 so G/N is an abelian simple group and must therefore be cyclic of order 2, 3 or 7.

Let

$$r = \begin{pmatrix} 0 & 0 & 1 \\ 1 & 0 & 1 \\ 0 & 1 & 0 \end{pmatrix} \in G \text{ and put } s = r^T r = \begin{pmatrix} 1 & 0 & 1 \\ 0 & 1 & 0 \\ 1 & 0 & 0 \end{pmatrix}$$

and a direct calculation reveals that $r \in G$ has order 7 and $s \in G$ has order 3. Therefore r^T also has order 7 but $r^T \notin \langle r \rangle$ so G has more than one Sylow 7-subgroup. Now G has no normal subgroup of index 2 since any group of order 84 contains a unique subgroup of order 7 by Sylow's theorem and so G would have only one Sylow 7-subgroup. Also N cannot have index 3 in G since then $r, r^T \in N$ so $s \in N$ which is absurd since 3 does not divide 56. It follows that $|G : N| = 7$. The group N contains all the Sylow 2-subgroups of G, of which there are 1 or 3. If there is a unique (normal) Sylow 2-subgroup S of G then G/S has order 21 and must have a unique Sylow 7-subgroup of index 3. It follows from the Correspondence Principle that G has a normal subgroup of index 3, a possibility which has already been shown the door. Therefore G has 3 Sylow 2-subgroups, the normalizer H of any one of which must have index 3 in G. Now H is not normal in G as we have previously established, so the natural action of G on $H \backslash G$ yields a homomorphism from G to S_3 with kernel of index more than 3. Thus S_3 is an epimorphic image of G. However, S_3 has a subgroup of index 2 and by the Correspondence Principle G must have a subgroup of index 2 which is impossible. Thus G is a simple group.

PSL$(2,7)$

We now move on to discuss $G = \mathrm{PSL}(2,7)$. Just as before we know that $|G| = 168$. When we blow away all the group theoretic froth from our discussion of $\mathrm{GL}(3,2)$, all that remains is that we are studying a group of order 168 and that it contains two elements of order 7 which generate a subgroup which contains an element of order 3. If we can find such elements in our new G, then we will be able to conclude that G is simple. If $a \in \mathrm{SL}(2,7)$ we write $[a]$ for the corresponding element of G.

We define some particular elements of G. Let

$$x = \left[\begin{pmatrix} 1 & 1 \\ 0 & 1 \end{pmatrix} \right] \text{ and } y = \left[\begin{pmatrix} 1 & 0 \\ 1 & 1 \end{pmatrix} \right],$$

so x and y are both of order 7, and we put

$$u = xyyx = \left[\begin{pmatrix} 3 & 4 \\ 2 & 3 \end{pmatrix} \right].$$

One can directly verify that the order of u is 3, so G is a simple group.

Now Proposition 4.14 applies and we deduce the remarkable result that $\mathrm{GL}(3,2)$ and $\mathrm{PSL}(2,7)$ are isomorphic groups.

4.2 Infinite Groups

4.2.1 Infinite Simple Groups

The finite groups of matrices and their factor groups which we have just been studying are baby versions of groups which arise naturally in geometry. In order to illustrate this important world, we will develop just enough linear algebra to make these groups accessible. Assume that V is a finite dimensional vector space over a field F. Some of what follows may remind the alert reader of a discussion concerning an infinite porcupine which followed Example 3.49. The group $GL(V)$ of all linear isomorphisms from V to V can be represented using matrices. Choose an ordered basis $v_1, \ldots, v_n$ for V, and to each $\alpha \in GL(V)$ you associate an $n \times n$ matrix a_{ij} via $\alpha(v_j) = \sum_i a_{ij} v_i$ for every $j = 1, \ldots, n$. This association is a group isomorphism from $GL(V)$ to $GL(n, F)$, the group of non-singular $n \times n$ matrices with entries in F. It really does not matter whether we use the abstract formulation $GL(V)$ or the concrete representation $GL(n, F)$. The first description is clean, and useful when reasoning about the group. The second description is better when you actually want to multiply (compose) elements. Let us stay abstract for a while. We have a map $det : GL(V) \to F^*$ which associates to each $\alpha \in GL(V)$ the field element which is the determinant of the matrix of α (with respect to any basis). As is well known, the determinant map is multiplicative: $\det \alpha\beta = \det \alpha \cdot \det \beta$ so the determinant map is a homomorphism of groups. In fact it is an epimorphism since the diagonal matrix $\mathrm{diag}(x, 1, \ldots, 1)$ has determinant x. The kernel of the determinant map is called the special linear group $SL(V)$ or in matrix terms $SL(n, F)$. Unless $n = 1$ the group $SL(V)$ is not the trivial group, and nor is it $GL(V)$ (unless the field has just two elements).

We investigate $SL(V)$. The first thing to do is to look at something which $SL(V)$ acts on, and since $SL(V) \le GL(V)$ and $GL(V)$ acts in a natural way on V, we have an action of $SL(V)$ on V. Suppose that $\gamma \in SL(V)$ is in the kernel of this action. Therefore $\gamma(v) = v$ for all $v \in V$, which means that γ is the identity map. Thus the kernel of this action is trivial. However, working a little harder we will be able to find an action with an interesting kernel.

Let Ω denote the set of one dimensional subspaces of V. Thus $P \in \Omega$ means that $P = \langle v \rangle$ for some non-zero $v \in V$. The action is as follows: if $\alpha \in GL(V)$, then $\alpha * \langle v \rangle = \langle \alpha(v) \rangle$. There is an issue over the legitimacy of this definition, but if $\langle v \rangle = \langle w \rangle$, then there is a non-zero scalar λ such that $v = \lambda w$, so $\alpha(v) = \lambda \alpha(w)$ by linearity, and so $\langle \alpha(v) \rangle = \langle \alpha(w) \rangle$. Having cleared away the bureaucracy, actually checking the action axioms is a triviality. Thus we have an action of $GL(V)$ on Ω, and therefore of $SL(V)$ on Ω. A linear transformation θ in one of these groups will act trivially if and only if for each $v \in V$ there is

$\lambda_v \in F$ such that $\theta(v) = \lambda_v v$.

Casually then, the effect of θ is to multiply vectors by scalars. We will now show that θ actually multiplies all vectors by the *same* scalar. If V is at most 1-dimensional, this is clear. If V is more than 1-dimensional, choose $v, w \in W$ which are linearly independent. Now $\theta(v + w) = \theta(v) + \theta(w)$ so $\lambda_{v+w}(v + w) = \lambda_v v + \lambda_w w$. Rearranging we have

$$(\lambda_{v+w} - \lambda_v)v + (\lambda_{v+w} - \lambda_w)w = 0.$$

Now linear independence forces $\lambda_v = \lambda_{v+w} = \lambda_w$. Now any $w \in W$ is either an element of $\langle v \rangle$ or v, w are linearly independent. Thus $\lambda_w = \lambda$ is independent of w, and the effect of θ is just scalar multiplication.

The kernels of the actions therefore consist of the scalar multiplications in each of the relevant groups. It is a fairly routine matter to verify that the centres of GL(V) and SL(V) are comprised of the scalar multiplications in each group. In matrix terms these scalar multiplications are scalar matrices (i.e. scalar multiples of the identity matrix).

This is sometimes an obstruction to the simplicity of SL(V). For example, working with matrices for notational ease, The scalar matrices in SL(2, $\mathbb{R}$) are $\pm I_2$, but the only scalar matrix in SL(3, $\mathbb{R}$) is I. The scalar matrices in SL(n, $\mathbb{C}$) are of the form ωI_n where ω is a complex n-th root of 1.

The group PSL(V) is obtained from SL(V) by factoring out the scalar matrices. Provided that V is at least 2-dimensional and the field is infinite, this will be an infinite non-abelian simple group. We omit the proof.

4.2.2 An Infinite Alternating Group

Let $S = \text{Sym } \mathbb{N}$. This group is not simple since its finitary elements form a normal subgroup (recall that $\alpha \in \text{Sym } \mathbb{N}$ is finitary if it moves only finitely many elements). Let S^f be the subgroup of finitary permutations. Every finitary permutation is the product of finitely many transpositions since we may as well be working in S_n for n sufficiently large. Moreover, it also follows that each element of S is either even or odd (and not both). This is again clear from working in S_n for n sufficiently large. The even elements form a normal subgroup A^f of index 2 in S^f.

Proposition 4.15

In the notation we have just established, the group A^f is an infinite non-abelian simple group.

Proof

Certainly A^f is infinite since it contains the 3-cycle $(m, m+1, m+2)$ for every $m \in \mathbb{N}$. It is non-abelian since restricting attention to those elements of A^f which fix every $m > 5$ yields a subgroup $A(5)$ which is a copy of A_5, and A_5 is non-abelian. Similarly by restricting attention to those elements of A^f which fix every $m > t$ yields a subgroup $A(t)$ which is a copy of A_t. Note that

$$1 = A(1) \leq A(2) \leq A(3) \cdots$$

and that

$$\bigcup_{t \in \mathbb{N}} A(t) = A^f.$$

Suppose for contradiction that K is a non-trivial normal subgroup of A^f. Choose any $k \in K - \{1\}$. Thus k has finite support so $k \in A(n)$ for all sufficiently large n. Thus $K \cap A(n) \neq 1$ for all sufficiently large n. Now $1 \neq K \cap A(n) \trianglelefteq A(n)$ and $A(n)$ is simple for all sufficiently large n. Therefore $A(n) \leq K \leq A^f$ for all sufficiently large n, say for $n \geq s$. Therefore

$$A^f = \bigcup_{n=s}^{\infty} A(n) \leq \bigcup_{n=s}^{\infty} K = K \leq A^f$$

so $A^f = K$ and we deduce that A^f is a simple group.

$\square$

We now back away from thinking about simple groups, and examine other interesting properties which infinite groups may have. The cardinality of the underlying set is clearly an important quantity.

Proposition 4.16

There is a countable abelian group which has uncountably many distinct subgroups (so it has uncountably more subgroups than elements!).

Proof

The countable group we will discuss is the additive group $\mathbb{Q}$ of rational numbers. Let π be a set of prime numbers. A π-number n is a natural number n such that all prime divisors of n are elements of π. Let $\mathbb{Q}_\pi \subseteq \mathbb{Q}$ be defined by $\mathbb{Q}_\pi = \{a/b \mid a \in \mathbb{Z}, b \in \mathbb{N}, \ b \text{ a } \pi\text{-number}\}$.

We now verify that each $\mathbb{Q}_\pi$ is an additive subgroup of $\mathbb{Q}$. For every π we have $0 = 0/1 \in \mathbb{Q}_\pi$ so $\mathbb{Q}_\pi \neq \emptyset$. Now if $a_1, a_2 \in \mathbb{Z}$ and $b_1, b_2 \in \mathbb{N}$ with each b_i

a π-number, then $a_1/b_1 - a_2/b_2 = (a_1 b_2 - a_2 b_1)/b_1 b_2 \in \mathbb{Q}_\pi$ because $b_1 b_2$ is a π-number. Thus each $\mathbb{Q}_\pi$ is an additive subgroup of $\mathbb{Q}$. Now if $\pi \neq \pi'$, then without loss of generality we may assume that there is $p \in \pi - \pi'$, so $1/p \in \mathbb{Q}_\pi$ but $1/p \notin \mathbb{Q}_{\pi'}$. Therefore $\mathbb{Q}_\pi \neq \mathbb{Q}'_\pi$. Now there are uncountably many subsets π of the prime numbers, and so uncountably many subgroups $\mathbb{Q}_\pi$ of $\mathbb{Q}$.

$\square$

Note that if $\alpha : G \to H$ is a group homomorphism and G is a finitely generated group, then Im α must also be finitely generated. Recall that if $G = \langle X \rangle$, then each element of G is expressible as a word (a product) using the elements of X and their inverses as letters. Thus each element of Im α is expressible as a word (a product) in the elements of $Y = \{(x)\alpha \mid x \in X\}$ and their inverses. If X is finite, then so is Y, so our result follows. In particular if two groups are isomorphic, and one of them is finitely generated, then so too is the other. The Cartesian product of finitely many finitely generated groups is clearly finitely generated.

Proposition 4.17

Let G be a group, and suppose that H is a subgroup of finite index in G. It follows that G is finitely generated if and only if H is finitely generated.

Proof

As is traditional, we dispose of the easier of the two implications first. Suppose that H is finitely generated by a set Y. There is a finite subset S of G such that $G = HS$. Now if $g \in G$, then $g = hs$ for some $h \in H$ and $s \in H$. It follows that $G = \langle Y \cup S \rangle$ and so G is finitely generated.

We begin again, and this time we suppose that G is generated by the finite set X. By replacing X by the finite set $X \cup \{x^{-1} \mid x \in X\}$, it is harmless to assume that X is closed under inversion. Choose a (finite) set of right coset representatives T for H in G such that the representative of H is 1.

For each $x \in X$ and $t \in T$ we have $tx \in Ht'$ for some $t' \in T$. We write $h_{x,t} = txt'^{-1} \in H$. Let $S = \{h_{x,t} \mid x \in X, t \in T\}$, a finite set. We shall show that $H = \langle S \rangle$.

Suppose that $h = x_1 x_2 \ldots x_n \in H$ with each $x_i \in X$. Define a finite sequence (t_i) of elements of T by $t_1 = 1$ and $t_i x_i \in Ht_{i+1}$ so $Ht_i x_i = Ht_{i+1}$ for each $i \leq n$. It follows that

$$Ht_{n+1} = Ht_n x_n = Ht_{n-1}x_{n-1}x_n = \cdots = Hx_1 \cdots x_n = Hh = H$$

so $t_{n+1} = 1$.

Now

$$h = t_1 x_1 (t_2^{-1} t_2) x_2 (t_3^{-1} t_3) \cdots (t_n^{-1} t_n) x_n t_{n+1}^{-1}$$
$$= (t_1 x_1 t_2^{-1}) \cdot (t_2 x_2 t_3^{-1}) \cdots (t_n x_n t_{n+1}^{-1})$$

which is a product of elements of S, and so we are done.

$\square$

Remark 4.18

An optimist might be tempted to hope that a subgroup of a finitely generated group must be finitely generated, but this is false. Consider the following pair of bijections α, β between $\mathbb{Z}$ and $\mathbb{Z}$. Let $(m)\alpha = m + 2$ for every $m \in \mathbb{Z}$. Let $(0)\beta = 1$ and $(1)\beta = 0$, and for every integer $n \notin \{0, 1\}$ we let $(n)\beta = n$. Let $G = \langle \alpha, \beta \rangle \leq \operatorname{Sym} \mathbb{Z}$ so G is finitely generated. Note that β^{α^i} swaps $2i$ with $2i + 1$ for every $i \in \mathbb{Z}$. Let $H = \langle \beta^{(\alpha^i)} \mid i \in \mathbb{Z} \rangle$. We will show that H is not a finitely generated group. Each $\beta^{(\alpha^i)}$ moves exactly two integers; accordingly we say that each $\beta^{(\alpha^i)}$ has support of size 2. Any product of elements of finite support must have finite support, so each element of H has finite support. Now suppose that A is a finite subset of H. It follows that the union U of all the supports of the elements of A is finite. Thus each element $a \in \langle A \rangle$ has support which is a subset of U. Now choose a natural number sufficiently large that $2i \notin U$. Now $\alpha^{(\beta^i)}$ moves $2i$ and so is not an element of $\langle A \rangle$, so $\langle A \rangle \neq H$. We deduce that H is not finitely generated because A was an arbitrary finite subset of H.

Definition 4.19

A group G is said to be *torsion-free* if the only element of G which has finite order is the identity.

The additive group of the real numbers is torsion-free, but the multiplicative group of non-zero real numbers is not torsion-free (because of -1). A group where every element has finite order is called a torsion group.

There is some rival language. A group G is said to be periodic if each element of G has finite order, and is aperiodic if only the identity element has finite order. Thus " periodic" means the same as "torsion" and "aperiodic" means the same as "torsion-free". Note that the infinite dihedral group is nether periodic nor aperiodic.

The following result is beautiful in its own right, and has a variety of illuminating proofs.

Theorem 4.20

Suppose that G is a torsion-free group containing an infinite cyclic group H with $|G : H| < \infty$. It follows that G is an infinite cyclic group.

Proof

Replacing H by its core, it is harmless to assume that $H \trianglelefteq G$ (Remark 3.37 ensures that $|G : \mathrm{core}_G(H)| < \infty$). We induct on $|G : H|$ with H an infinite cyclic normal subgroup of a torsion-free group G. If $|G : H| = 1$, then $G = H$ and so G is an infinite cyclic group.

Suppose that there is N such that $H \triangleleft N \triangleleft G$, so N is a normal subgroup of G strictly between H and G. In this event $|N : H| < |G : H|$. Now N is a subgroup of a torsion-free group and so is torsion-free, so N is an infinite cyclic group by inductive hypothesis. Now $|G : N| < |G : H|$ and N is infinite cyclic. Once again by inductive hypothesis we may deduce that G is infinite cyclic.

Thus the proof is complete unless there is no normal subgroup of G strictly intermediate between H and G, in which event $M = G/H$ is a finite simple group by the Correspondence Principle, or the third isomorphism theorem if you prefer. Note that any proper subgroup of M must be of the form T/H where T is a proper subgroup of G. Thus T is torsion-free and by inductive hypothesis T is an infinite cyclic group. Now T/H is a homomorphic image of a cyclic group and so must be cyclic. We conclude that M is a finite simple group with all its proper subgroups being cyclic. There are various ways to clinch the proof from here; this particular variation was suggested by Gunnar Traustason.

First we assume that M is non-cyclic so M has at least two different maximal subgroups X and Y. (If X were the only maximal subgroup, we could choose $g \in M - X$ and then there would be no maximal subgroup containing $\langle g \rangle$. This forces $\langle g \rangle = M$.)

Now $X \cap Y$ is normalized by $\langle X, Y \rangle = M$ so $X \cap Y = 1$. Since M is non-abelian we have $X \neq 1$ and so $N_M(X) = X$, otherwise the maximality of X would force it to be a non-trivial normal subgroup of the simple group M. Now X has $|M : X|$ conjugates containing a total of $|M : X|(|X| - 1) = |M| - |M : X|$ non-identity elements. Note that the union of the conjugates of X is not M, so we may choose a maximal subgroup Y which is not a conjugate of X. The union of the conjugates of Y contains $|M| - |M : Y|$ non-identity elements, none of which are in a conjugate of X. Thus

$$|M| - |M : X| + |M| - |M : Y| + 1 \leq |M|$$

so

$$|M| + 1 \leq |M : X| + |M : Y|.$$

Now $|M : X|$, $|M : Y| \leq |M|/2$ so $|M| + 1 \leq |M|$ which is absurd. It follows that M must be a cyclic simple group and so of prime order p.

Recall that $H = \langle h \rangle$ is infinite cyclic. Choose $s \in G - H$ and put $S = \langle s \rangle$. Now $HS = G$ so $S/H \cap S \simeq HS/H$ is finite. It follows that $H \cap S$ is infinite, so $H \cap S = \langle h^j \rangle$ where $j > 0$. Observe that $h^s = h^{\pm 1}$ and $(h^j)^s = h^j$ since $h^j \in \langle s \rangle$. However h has infinite order so $h^j \neq h^{-j}$. Therefore $h^s = h$ and G is an abelian group. Now G is torsion-free abelian and contains a infinite cyclic subgroup of finite index m. The map defined by $g \mapsto g^m$ is a monomorphism and so G is an infinite subgroup of an infinite cyclic group. Therefore G is an infinite cyclic group by Proposition 1.45.

□

That was quite a lot of work. However, we used only very elementary notions. We will now present an idea which yields a simpler proof.

4.2.3 The Transfer Map

Suppose that G is a group with an abelian subgroup A of finite index. Choose a right transversal (t_i) for A where $1, \ldots, n$ and $n = |G : A|$. Suppose that $g \in G$. For each i we have $t_i g = a_{i,g} t_{i'}$ for uniquely determined i' and $a_{i,g} \in A$. Let $(g)\tau = \prod_{i=1}^{n} a_{i,g}$ and the order of the product does not matter because A is abelian. Thus

$$(g)\tau = \prod_{i=1}^{n} t_i g t_{i'}^{-1}.$$

Now to get some cancelling going, it makes sense to choose the order in which the $a_{i,g}$ are being multiplied to coincide with the order in which $1, \ldots, n$ appear in the cycle shape of the permutation $i \mapsto i'$. Suppose there are m disjoint cycles and choose a representative m_j from each cycle, and suppose that cycle has length l_j for each j in the range $1 \leq j \leq m$. Now

$$(g)\tau = \prod_{j=1}^{m} t_{m_j} g^{l_j} t_{m_j}^{-1}$$

so $(g)\tau$ is a product of conjugates of powers of g. Moreover, each $t_{m_j} g^{l_j} t_{m_j}^{-1} \in A$. It is now clear that the transfer is independent of the chosen coset representatives t_i since if $\widehat{t_i}$ is a rival choice, then $a_i t_i = \widehat{t_i}$ for each i, so

$$\widehat{t_{m_j}} g^{l_j} \widehat{t_{m_j}}^{-1} = a_{m_j} t_{m_j} g^{l_j} t_{m_j}^{-1} a_{m_j}^{-1} = t_{m_j} g^{l_j} t_{m_j}^{-1},$$

where the final equality holds because A is abelian and both $a_{m_j}, t_{m_j} g^{l_j} t_{m_j}^{-1} \in A$ for each j.

Now we show that $\tau : G \to A$ is a homomorphism of groups. To see this, we return to the definition of τ and apply it to a product xy where $x, y \in G$. We will have to be a little more careful about the permutation $i \mapsto i'$ of the subscripts arising from right multiplication by g. We will now refer to this permutation as π_g. Now

$$(xy)\tau = \prod_{i=1}^{n} t_i xy t_{(i)\pi_{xy}}^{-1} = \prod_{i=1}^{n} t_i x t_{(i)\pi_x}^{-1} t_{(i)\pi_x} y t_{(i)\pi_{xy}}^{-1}$$

$$= (\prod_{i=1}^{n} t_i x t_{(i)\pi_x}^{-1})(\prod_{k=1}^{n} t_k x t_{(k)\pi_y}^{-1}) = (x)\tau \cdot (y)\tau.$$

Thus the transfer map is a homomorphism.

Proposition 4.21

Suppose that G is a group with a central subgroup Z of finite index n in G, then the transfer map $\tau : G \to Z$ has the property that $(g)\tau = g^n$ for every $g \in G$.

Proof

The formula we have worked out yields $(g)\tau = \prod_{j=1}^{m} t_{m_j} g^{l_j} t_{m_j}^{-1}$ where for each j the element $t_{m_j} g^{l_j} t_{m_j}^{-1}$ is in $Z(G)$. Therefore $t_{m_j} g^{l_j} t_{m_j}^{-1} = t_{m_j}^{-1} t_{m_j} g^{l_j} t_{m_j}^{-1} t_{m_j} = g^{l_j}$ and so $(g)\tau = \prod_{j=1}^{m} g^{l_j} = g^n$.

$\square$

This is outrageous at first sight. After all, it is clear that $g^n \in Z$ for each $g \in G$, but it is far from clear that $(xy)^n = x^n y^n$ for each $x, y \in G$. However, raising elements to the power n is how you apply the transfer map in this context, and we have proved that the transfer map is a homomorphism so indeed $(xy)^n = x^n y^n$ for each $x, y \in G$.

In order to give an insight into the potency of this map, we introduce a light dose of some theory which will be developed more seriously in Chapter 5.

4.3 The Derived Group

Definition 4.22

Suppose that G is a group. Let $\mathfrak{A}$ (a gothic "A") denote the set of normal subgroups N of G which have the property that G/N is abelian. We call $G' = \cap_{N \in \mathfrak{A}} N$ the derived group of G.

Proposition 4.23

Suppose that G is a group with derived subgroup G'. It follows that $G' \trianglelefteq G$, G/G' is abelian and
$$G' = \langle g^{-1}h^{-1}gh \mid g, h \in G \rangle.$$

Proof

Since G' is an intersection of normal subgroups of G, it follows that G' is a normal subgroup by the solution to Exercise 2.9, p.52. Now suppose that $N \trianglelefteq G$ and G/N is abelian. Take any $g, h \in G$, then

$$Ng^{-1}h^{-1}gh = Ng^{-1}Nh^{-1}NgNh = Ng^{-1}NgNh^{-1}Nh = Ng^{-1}gh^{-1}h = N$$

so $g^{-1}h^{-1}gh \in N$. Therefore each $g^{-1}h^{-1}gh \in G'$. Thus $g^{-1}h^{-1}ghG' = G'$ so $ghG' = hgG'$ and therefore $G'gh = G'hg$. We now have $G'gG'h = G'hG'g$ but $g, h \in G$ were chosen arbitrarily so G/G' is an abelian group.

Let $X = \langle g^{-1}h^{-1}gh \mid g, h \in G \rangle$. As a by-product of the first part of our proof we obtain that $\{g^{-1}h^{-1}gh \mid g, h \in G\} \subseteq G'$ and so $X \leq G'$. Next we show that $X \trianglelefteq G$. Suppose that $z \in G$, then

$$\begin{aligned}
\{g^{-1}h^{-1}gh \mid g, h \in G\}^z &= \{(g^z)^{-1}(h^z)^{-1}g^z h^z \mid g, h \in G\} \\
&= \{(g^{-1})h^{-1}gh \mid g, h \in G\}.
\end{aligned}$$

Since any $z \in G$ conjugates a generating set of X to itself, it follows that $z^{-1}Xz = X$ for all $z \in G$. Therefore $X \trianglelefteq G$. Moreover, since every element of the form $g^{-1}h^{-1}gh$ is in X, we can recycle the argument which showed that G/G' is abelian to show that G/X is abelian. Now the definition of G' forces $G' \leq X$ and therefore $G' = X$.

$\square$

If G is an abelian group, then $G' = 1$. If G is a non-abelian simple group, then $G' = G$. Recall from the remarks following Definition 2.36 that elements of the form $g^{-1}h^{-1}gh$ are called commutators. Thus G' is the subgroup of G generated by the commutators.

An immediate application is a celebrated result of the great mathematician Issai Schur.

Proposition 4.24 (Schur)

Suppose that G is a group and that its centre $Z(G)$ is of finite index n in G. It follows that G' is finite.

Proof

Choose a right transversal $x_1, \ldots, x_n$ for $Z(G)$ in G. Suppose that $a, b \in G$ so $a = wx_i$ and $b = zx_j$ for some $w, z \in Z(G)$ and $1 \le i, j \le n$. Now

$$a^{-1}b^{-1}ab = (wx_i)^{-1}(zx_j)^{-1}wx_izx_j = x_i^{-1}w^{-1}x_j^{-1}z^{-1}wx_izx_j.$$

The elements w, z are central so

$$a^{-1}b^{-1}ab = w^{-1}wz^{-1}zx_i^{-1}x_j^{-1}x_ix_j = x_i^{-1}x_j^{-1}x_ix_j.$$

There are only finitely many commutators, so G' is a finitely generated group. Also $|G : Z(G)|$ is finite so $G'/G' \cap Z(G) \simeq G'Z(G)/Z(G)$ is finite. We have used the second isomorphism theorem. Now $G' \cap Z(G)$ is of finite index in G' which is a finitely generated group. Therefore $G' \cap Z(G)$ is finitely generated by Proposition 4.17.

We now apply the transfer map τ from G to $Z(G)$, recalling that $|G : Z(G)| = n$. Notice that $G' \le \operatorname{Ker} \tau$ since $Z(G)$ is abelian. If $t \in G'$, then $(t)\tau = 1 = t^n$. The group $G' \cap Z(G)$ is a finitely generated abelian group and the order of each of its elements is finite. We can now apply our knowledge of the structure of finitely generated abelian groups [Theorem 2.44] (or reflect for a moment or two) to deduce that $G' \cap Z(G)$ is a finite group. Now G' has a finite subgroup of finite index, and so is finite itself.

$\square$

Let us reflect on Schur's remarkable result. It says that if a group has a large central subgroup, then it has a large abelian homomorphic image (because G/G' is abelian and G' is only finite). We are currently doing that kind of group theory where we think of finite groups as pathetic little objects, annoying little edge-effects which are dwarfed by their infinite cousins. This contrasts very much with Chapter 3 for example, where the combinatorics of finite groups (especially finite p-groups) was dominant. It is important to be able to shift your point of view to suit the problem in hand.

We are now in a position to revisit Theorem 4.20. We quickly restate the result: suppose that G is a torsion-free group containing an infinite cyclic group H with $|G : H| < \infty$. It follows that G is an infinite cyclic group.

Proof

(Mark II) Since G has a finitely generated (cyclic) subgroup of finite index, it follows that G is finitely generated. Take any non-identity element $g \in G$ then $\langle g \rangle$ is an infinite group since G is torsion-free. It follows by Proposition 2.30 that $\langle g \rangle \cap H$ has finite index in H and therefore in G. However $\langle g \rangle \leq C_G(g)$ so we deduce that all centralizers of elements of G have finite index in G. Now G is finitely generated by X say. Now $z \in G$ is central if and only if it commutes with each of the finitely many elements of X. Thus

$$Z(G) = \bigcap_{x \in X} C_G(x).$$

Now the intersection of finitely many subgroups of finite index in G must have finite index in G by Proposition 1.32. Thus $|G : Z(G)|$ is finite. Now G' is finite by Schur's Proposition 4.24, but G is torsion-free so $G' = 1$ and therefore G is abelian. The map which sends each g to $g^{|G:H|}$ is a monomorphism from G to a non-trivial subgroup of H. The reason that it is injective is that the kernel must be trivial since G is torsion-free.

We know that all non-trivial subgroups of H are infinite cyclic by Proposition 1.45 so G is infinite cyclic.

$\square$

While we are blundering around in this particular playpen, it is hard to resist mentioning the next result.

Proposition 4.25 (Dicman's lemma)

Let G be a group generated by finitely many elements each of which has finite order and only finitely many conjugates. It follows that G is a finite group.

Proof

Let the generating elements in question form a set X. Now $|G : C_G(x)|$ is finite for each $x \in X$ since x belongs to a finite conjugacy class. Observe that $Z(G) = \cap_{x \in X} C_G(x)$ since X generates G. Now by Proposition 1.32, the intersection of finitely many subgroups of finite index is itself of finite index, so $|G : Z(G)| < \infty$. We deduce from Proposition 4.24 that G' is a finite group. However, G/G' is a finitely generated group (it is a homomorphic image of the finitely generated group G), is abelian, and is generated by elements of finite order. It follows that G/G' is a finite group. Thus $|G| = |G : G'||G'|$ is finite.

$\square$

5.1 The Commutator Calculus

Suppose that G is a group and that $x, y \in G$. We write $[x, y]$ for the element $x^{-1}y^{-1}xy$. This element $[x, y]$ is called the *commutator* of x and y. We have briefly alluded to commutators after Definition 2.36 and concerning Schur's Proposition 4.24, but now we make a more serious study of these objects, following the giant footsteps of Philip Hall.

Notice that $xy = yx[x, y]$ so $[x, y]$ measures the failure of x and y to commute. We make a start by examining some of the ways in which commutators interacts with multiplication, inversion, conjugation and other commutators. The notation $[a, b, c]$ means $[[a, b], c]$. We will assemble a calculus of commutators; a collection of relatively simple rules which will enable us to manipulate complicated expressions. This is algebra proper.

Proposition 5.1

Suppose $x, y, z \in G$ where G is any group. The following equations all hold:

(a) $[y, x] = [x, y]^{-1}$.

(b) $[x, y]^z = [x^z, y^z]$.

(c) $[xy, z] = [x, z]^y [y, z]$.

(d) $[x, yz] = [x, z][x, y]^z$.

(e) $[x, yz] = [x, z][x, y][x, y, z]$.

Proof

In each case you have the option of simply writing out each side of the equation separately and in full without the use of commutator symbols, and then performing any cancelling that is available. You will then notice that you get the same answer from each of the two calculations, and the proof of that equation will be complete. That is the cheap and nasty method of great utility under examination conditions, or when all else fails. We will instead give more conceptual arguments which begin to give an explanation as to *why* the formulas are valid.

(a) $xy = yx[x,y] = xy[y,x][x,y]$ so $[y,x] = [x,y]^{-1}$.

(b) $xy = yx[x,y]$ so $x^z y^z = y^z x^z [x,y]^z$. However, we also have $x^z y^z = y^z x^z [x^z, y^z]$ and so $[x,y]^z = [x^z, y^z]$.

(c)

$$
\begin{aligned}
xyz &= (xy)z = z(xy)[xy,z] = (zx)y[xy,z] = (xz[z,x])y[xy,z] \\
&= xzy(y^{-1}[z,x]y)[xy,z] = x(zy)[z,x]^y[xy,z] \\
&= x(yz[z,y])[z,x]^y[xy,z] = xyz[z,y][z,x]^y[xy,z]
\end{aligned}
$$

so $[z,y][z,x]^y[xy,z] = 1$ and therefore $[xy,z] = [x,z]^y[y,z]$. Thus fixing the second entry of the commutator to be z and regarding commutation as a function $[-,z]$ of the first entry is not quite a homomorphism. It almost is one though, but conjugation by y spoils the party.

(d) $[x,yz] = [yz,x]^{-1}$ by part (a). Thus $[x,yz] = ([y,x]^z[z,x])^{-1} = [x,z][x,y]^z$ using parts (a) and (c). Thus fixing the first entry of the commutator to be x and regarding commutation as a function $[x,-]$ of the second entry is also not quite a homomorphism. Not only is there conjugation by z to mess things up, but also the product occurs the wrong way round. Thus (c) looks a touch better than (d).

(e)

$$
xyz = yx[x,y]z = yxz[x,y][x,y,z] = yzx[x,z][x,y][x,y,z]
$$

whereas $xyz = yzx[x,yz]$ so $[x,yz] = [x,z][x,y][x,y,z]$.

$\square$

EXERCISES

5.1 Suppose that G is a group and that $x,y,z \in G$. Prove that

$$
[xy,z] = [x,z][x,z,y][y,z].
$$

5.2 Suppose that G is a group and that $x, y, z \in G$. Prove that

$$[z, xy][y, zx][x, yz] = 1.$$

5.3 Suppose that G is a group and that $x, y, z \in G$. Prove the Hall–Witt identity:

$$[x, y^{-1}, z]^y [y, z^{-1}, x]^z [z, x^{-1}, y]^x = 1.$$

5.4 Suppose that the group G has a subgroup H. Prove that $H \trianglelefteq G$ if and only if $[h, g] \in H$ for each $h \in H$ and for each $g \in G$.

5.5 In the group S_3, which elements arise as commutators?

5.6 Let G be a group. Suppose that $\varphi : G \to H$ is a group epimorphism. Let $C = \{[x, y] \mid x, y \in G\}$. Show that H is abelian if and only if $C \subseteq \operatorname{Ker} \varphi$.

5.2 The Derived Group Revisited

We discussed the derived group briefly in Section 4.3. If you have not read that section, it is not logically necessary to read it now (indeed, we put another spin on things this time, giving a different but equivalent definition of the derived group). However, the reader may wish to refer back to the earlier remarks to add depth to his or her understanding. There is a small overlap between material in Chapter 4 and the exposition here.

Definition 5.2

Let G be a group, and put

$$G' = \langle [x, y] \mid x, y \in G \rangle$$

so G' is a subgroup of G. It is variously called the *derived group* of G or the *commutator subgroup* of G.

The solution to Exercise 5.6 shows that G' is a subgroup of every normal subgroup N of G which has the property that G/N is abelian (use the natural epimorphisms $\pi : G \to G/N$). On the other hand, the conjugate of a commutator is a commutator thanks to Proposition 5.1 (b), so our set of generators for G' is closed under conjugation and therefore $G' \trianglelefteq G$. Moreover for each $x, y \in G$ we have $[G'x, G'y] = G'[x, y] = G'$ so G/G' is an abelian group. Thus G' can be characterized as the intersection of the collection of normal subgroups N of G which have the property that G/N is abelian, and moreover G' is the unique smallest subgroup in this collection of subgroups. Also notice that if $G' \leq H \leq G$ then G/G' is abelian, so $H/G' \trianglelefteq G/G'$ and so by the Correspondence Principle $H \trianglelefteq G$.

The minimality of G' among normal subgroups of G with abelian quotients means that in some sense G/G' must be a maximal abelian quotient of G. We will now try to bring that fuzzy remark a little more into focus.

Proposition 5.3

Let G be a group and suppose that π is the natural epimorphism $G \to G/G'$. Suppose that $\theta : G \to H$ is a group homomorphism with an abelian image. It follows that there is a unique group homomorphism $\widehat{\theta} : G/G' \to H$ such that $\pi \circ \widehat{\theta} = \theta$.

Proof

Since Im θ is abelian, it follows that $G' \leq$ Ker θ. Now the useful form of the third isomorphism theorem applies (Theorem 2.32) so the homomorphism $\widehat{\theta}$ exists and is unique. The way the proof goes, it is not surprising how $\widehat{\theta}$ is defined. For each $x \in G$ we have $(G'x)\widehat{\theta} = (x)\theta$.

□

This group G/G' is clearly important, and so has attracted a multiplicity of notation and names. It is written $ab(G), \text{Ab}(G)$ or G^{ab}. We favour the last of these alternatives. The group is called by the following names: the *abelianization of G*, the *commutator quotient of G* and *G made abelian*. The repellent term "abelianization" is formed by turning a proper name into an adjective, which then becomes a verb, and finally is rendered a noun once more.

If G is finitely generated, then so is every factor group of G. Thus G^{ab} is finitely generated. There is a structure theorem for finitely generated abelian groups (Theorem 2.44), so G^{ab} is subject to that result. If G^{ab} is infinite, then G must be infinite. However, it may be that G^{ab} is finite but G is infinite. A witness to this fact is the infinite dihedral group $D \simeq D_\infty$ (see Definition 1.48) which is a non-abelian group with an infinite cyclic group $\langle z \rangle$ of index 2. The reader is urged to check that every element of $\langle z^2 \rangle$ is a commutator, and so $\langle z^2 \rangle \leq D'$. However, it is a straightforward matter to verify that $D/\langle z^2 \rangle \simeq C_2 \times C_2$, an abelian group. Thus $D' \leq \langle z^2 \rangle$ and so $D' = \langle z^2 \rangle$ and $|D : D'| = 4$.

Much of the previous discussion does not depend crucially on the particular form of the commutator $[x, y] = x^{-1}y^{-1}xy$. Any word would do. For example, the word x^2 could play the role of $[x, y]$. In this case if G is a group, then we can define $G^2 = \langle g^2 \mid g \in G \rangle$. We must be very careful with this notation, because we have used exactly the same symbolism to denote the Cartesian square $G \times G$ of G. These notations will certainly not mix, but we have no need of Cartesian squares in the immediate future so we will take a chance. Now G^2 will be the intersection of all the normal subgroups M of G which have the property that each element of G/M becomes the identity element when you square it. Moreover G/G^2 is abelian, since the map $\theta : G/G^2 \to G/G^2$ defined by $(x)\theta = x^2$ for every $x \in G/G^2$ has trivial image, and is therefore a homomorphism. The solution to Exercise 2.7, p.45 applies. Alternatively you can supply a direct proof.

Proposition 5.4

Let G be a group and suppose that π is the natural epimorphism $G \to G/G^2$. Suppose also that $\theta : G \to H$ is a group homomorphism with the property that $y^2 = 1$ for every $y \in \operatorname{Im} \theta$. It follows that there is a unique group homomorphism $\widehat{\theta} : G/G^2 \to H$ such that $\pi \circ \widehat{\theta} = \theta$.

Proof

The proof is in exactly the same spirit as that of Proposition 5.3.

$\square$

Definition 5.5

Let n be a natural number. We say that a group G has exponent n if $g^n = 1$ for every $g \in G$.

Remark 5.6

Sometimes people define the exponent m of a group G to be the least natural number (if there is one) such that $g^m = 1$ for every $g \in G$. Using this definition of exponent, a group H of exponent 2 does not have exponent 4. Using our official definition, it does! The problem is very similar to the one which arises when studying periodic functions. Since the function from $\mathbb{R}$ to $\mathbb{R}$ defined by $x \mapsto \sin 2x$ has period π, do we allow ourselves to say that $\sin 2x$ has period 2π? We say yes, and so we make Definition 5.5 accordingly. This has the advantage that the subgroups and factor groups of a group of exponent n all have exponent n. If we really want to emphasize that the exponent of a group G is n and no less, we will have to say that G has exponent exactly n.

We can play the game surrounding Propositions 5.3 and 5.4 using the word x^n where n is any fixed natural number. The subgroup of G defined by this word is $G^n = \langle g^n \mid g \in G \rangle$ and the analogue of Proposition 5.3 will hold when considering homomorphisms from G with images having exponent n.

Question 5.7

Does there exist a group G which can be generated by two elements and which has the property that G/G^5 is an infinite group? Put another way, is there an infinite 2-generator group of exponent 5?

At the time of writing this question is open (i.e. no-one on this world has published the answer).

Definition 5.8

Suppose that G is a group with subgroups

$$1 = G_0 \le G_1 \le G_2 \le \ldots \le G_n = G \quad (n \in \mathbb{N}) \tag{5.1}$$

We say that such a list of inclusions is a *series* for G. This series (5.1) is a *normal series* if each $G_i \trianglelefteq G$. If we have the weaker condition that each $G_{i-1} \trianglelefteq G_i$, the series is a *subnormal series*. The factors of a subnormal series are the groups G_i/G_{i-1} for $1 \le i \le n$ and the *length* of the series is n.

Example 5.9

Let $G = S_3$, the symmetric group on $\{1, 2, 3\}$. Now G has a normal (cyclic) subgroup N of order 3. Now $1 \le N \le G$ is a normal series for G of length 2. The two factors of the series are (up to isomorphism) C_3 and C_2.

A *group property* is an attribute a group G may have, with the property that if G has the attribute, and H is isomorphic to G, then H has the attribute. Examples of group properties are being finite, infinite, abelian, cyclic and simple. Examples of attributes which are not group properties include being a group with elements which are matrices, being a subgroup of a specified group, and being written down using green ink.

Let Q be a group property. If a group G has the property Q we may say that G is a Q-group. We are simply using the property as an adjective in a familiar way (if G has the property that it is finite, we may say that G is a finite group).

A group G is said to be a poly-Q-group if G has a subnormal series with each factor being a group with property Q. Now being a poly-Q-group is a (possibly new) property of groups. Notice that any Q-group is a poly-Q-group.

A group is polyfinite if and only if it is finite, which is why the term is hardly ever used. Polycyclic groups and polyabelian groups are classes of groups strictly larger than both cyclic groups and abelian groups, since S_3 is a witness to both of these facts. You might at first think that polyinfinite means the same as infinite, but that is not quite true. The trivial group has a subnormal series of length 0, and since there are no factors of this series, the factor groups all have any property you like! Trivial groups are the only polyinfinite groups which are finite of course.

The class of polyabelian groups is extremely important, and is given a special name.

Definition 5.10

We say that a group is *soluble* or *solvable* if it is polyabelian.

The terminology differs between British and North American usage. Some group theorists pretend to think that a soluble group is one which would disappear if you dropped it in your coffee.

Suppose that $A \lhd B$. Recall that B/A is a simple group if and only if there is no normal subgroup N of A such that $A < N < B$ (by the Correspondence Principle or the third isomorphism theorem). This indicates that there is a connection between simple groups and subnormal series. We shall prove later that any finite group is polysimple, and that the simple factors and their multiplicities are determined by G. Many questions about finite groups are thereby reduced to questions about simple groups. The classification of finite simple groups has therefore disposed of very many previously open questions about finite groups. However, it does not seem to be the case that every question about finite groups is amenable to this treatment, and finite group theory has survived the classification as a research discipline. However, compared to the glory days of the sixties and seventies, it exists in much reduced form. Reports of the death of finite group theory have been much exaggerated.

Definition 5.11

Suppose that
$$1 = G_0 \leq \ldots \leq G_n = G \quad (n \in \mathbb{N})$$
is a subnormal series for G. A *refinement* of this series is any subnormal series which can be obtained by inserting a finite number of extra subgroups in the

series.

For example, a trivial way of refining a series is to insert an extra copy of one of the groups in the series. It is much more interesting to refine a subnormal series by finding $G_{i-1} \lhd N \lhd G_i$ and inserting it in the series. Such an N will exists precisely when the factor G_i/G_{i-1} is non-trivial, but is not a simple group. Thus a subnormal series admits only trivial refinements exactly when each non-trivial factor is a simple group.

Definition 5.12

Suppose that G is a group equipped with two subnormal series

$$1 = G_0 \leq \ldots \leq G_n = G$$

and

$$1 = H_0 \leq \ldots \leq H_m = G.$$

We say that these series are *equivalent* if $n = m$, and there is a bijection $\sigma : \{1, \ldots, n\} \to \{1, \ldots, n\}$ such that $G_i/G_{i-1} \simeq H_{(i)\sigma}/H_{(i)\sigma-1}$ for every i in the range $1 \leq i \leq n$.

Theorem 5.13 (Jordan–Hölder Theorem)

Suppose that G is a group equipped with two subnormal series

$$1 = G_0 \leq \ldots \leq G_n = G$$

and

$$1 = H_0 \leq \ldots \leq H_m = G.$$

It follows that these series have equivalent refinements.

Proof

We insert groups between G_{i-1} and G_i for every i as follows:

$$G_{i-1} = G_{i-1}(H_0 \cap G_i) \unlhd G_{i-1}(H_1 \cap G_i) \unlhd \cdots$$

$$\unlhd G_{i-1}(H_j \cap G_i) \unlhd \cdots \unlhd G_{i-1}(H_m \cap G_i) = G_{i-1}G_i = G_i$$

and we keep track of the groups by putting $G_{i,j} = G_{i-1}(H_j \cap G_i)$. Thus for each i in the range $1 \leq i \leq n$ we have

$$G_{i-1} = G_{i0} \unlhd G_{i,1} \unlhd \cdots \unlhd G_{i,j} \unlhd \cdots \unlhd G_{i,m} = G_i.$$

Now we reverse the roles of the two series. We insinuate groups $H_{j,i}$ between H_{j-1} and H_j where $H_{j,i} = H_{j-1}(G_i \cap H_j)$ and so

$$H_{j-1} = H_{j,0} \trianglelefteq H_{j,1} \trianglelefteq \cdots \trianglelefteq H_{j,i} \trianglelefteq \cdots \trianglelefteq H_{j,n} = H_j.$$

We claim that

$$G_{i,j}/G_{i,j-1} \simeq H_{j,i}/H_{j,i-1}$$

for all legal values of i and j. However, this is an instance of Zassenhaus's Butterfly Lemma 2.34, and provides the required bijection.

$\square$

If a group G has a subnormal series with simple factors, then this series admits of only trivial refinements (repetitions). The isomorphism types of these factors (and their multiplicities) are independent of the subnormal series, and are therefore invariants of G (i.e. up to isomorphism, and the order in which they appear, these factors depend only upon G).

Definition 5.14

A subnormal series with simple factors is called a *composition series*, and the factors are called *composition factors*.

Any finite group has a composition series. The trivial group has a composition series of length 0 with no composition factors. If G is a non-trivial finite group we begin with the subnormal series $1 = G_0 \triangleleft G_1 = G$ and successively refine it in a non-trivial way (no repetitions) until this process can no longer be continued. At all times the order of G is the product of the orders of the factors of the series, and these must be natural numbers which are at least 2, and so after finitely many refinements the process must stop. At this stage we have a composition series for G.

Note that the composition factors of a finite abelian group must all be abelian, and therefore be cyclic of prime order. If n is a natural number and we list the composition factors of the cyclic group C_n, we obtain cyclic groups of prime order. For each prime p which divides n the factor C_p occurs in the composition series with the same multiplicity as the power of p which divides n. Thus C_{12} has composition factors C_2, C_2 and C_3 just as $12 = 2^2 \cdot 3$. The fact that the composition factors and their multiplicities are determined by the finite group G is therefore a generalization of Gauss's *Fundamental Theorem of Arithmetic* which asserts that every natural number n may be uniquely expressed as a product of prime numbers (with 1 being the empty product and the order of the primes forming the product being irrelevant).

Note that all finite abelian groups of the same order have the same composition factors with the same multiplicities, so the composition factors of G do not determine G up to isomorphism, as testified by C_4 and $C_2 \times C_2$.

Now let us think about infinite groups. An infinite simple group S does have a composition series. In fact

$$1 = S_0 \lhd S_1 = S$$

is its only composition series. However, not all infinite groups have composition series. For example, the infinite cyclic group C_∞ has no composition series. Any finite series for C_∞ must have cyclic factors (they are factor groups of subgroups of a cyclic group, and are therefore cyclic). At least one of the factors must be infinite cyclic since C_∞ is infinite. Now this infinite cyclic factor group is not simple so the series cannot be a composition series.

A polycyclic group group has a subnormal series where all factors are cyclic groups. Refining such a series does not change the number of infinite cyclic factors, because if H is infinite cyclic, and $K \lhd H$, then either $K = 1$ and H/K is infinite cyclic, or K is infinite cyclic with G/K finite cyclic. Thus the Jordan–Hölder Theorem 5.13 applies, and the number of infinite cyclic factors in a polycyclic series of a group G is independent of the polycyclic series. This is an invariant of G, and is often called the *Hirsch length* (or *Hirsch number*) of G after Kurt Hirsch, one of the pioneers of infinite group theory. Some of the more elementary properties of polycyclic groups can be obtained by induction on Hirsch length.

The notion of Hirsch length has immediate application to the theory of finitely generated abelian groups. In particular, it follows that any subgroup of finite index in $\oplus_{i=1}^n \mathbb{Z}^n$ is a copy of $\oplus_{i=1}^n \mathbb{Z}^n$.

5.3 Nilpotent Groups

Definition 5.15

Let G be a group. A normal series

$$1 = G_0 \leq G_1 \leq G_2 \leq \ldots \leq G_n = G$$

is called a *central series* if for each $i < n$ we have that $G_{i+1}/G_i \leq Z(G/G_i)$. We say that a group is *nilpotent* if it has a central series, and in this event we say that the least number of factors in a central series for G is the *nilpotency class* (or just the *class*) of G.

This condition that $G_{i+1}/G_i \leq Z(G/G_i)$ is equivalent to the commutator condition that $[G_i x, G_i g] = G_i$ for each $x \in G_{i+1}$ and for each $g \in G$. However,

$$[G_i x, G_i g] = G_i x^{-1} G_i g^{-1} G_i x G_i g = G_i[x, g]$$

so the condition can be simply restated as $[x, g] \in G_i$ for each i, for each $x \in G_{i+1}$ and for each $g \in G$. Informally and in words then, whenever you take a commutator of an element of G_{i+1} with an arbitrary group element, you end up in G_i.

The trivial group has nilpotency class 0, and non-trivial abelian groups have nilpotency class 1. In fact it is common usage to say that G has nilpotency class c to mean that the nilpotency class of G is at most c. This is sloppy but convenient, and enables you to say that abelian groups have nilpotency class 1 without having to mutter "except for the trivial group" under your breath. This is the same problem, with the same solution, as the exponent n issue we discussed earlier. If we wish to assert that G has nilpotency class c but does not have nilpotency class $c - 1$, then we will say that the nilpotency class of G is exactly c.

Proposition 5.16

Let P be a finite p-group. It follows that P is nilpotent.

Proof

We induct on the group order. The trivial group is nilpotent of class 0 so we are on our way. Thus we may suppose that $|P| > 1$. Therefore the centre $Z(P)$ of P is non-trivial by Proposition 3.38. The finite p-group $Q = P/Z(P)$ has smaller order than P, and so by induction is nilpotent. Thus there is a central series

$$1 = Q_1 \leq Q_2 \leq \ldots \leq Q_n = Q \qquad (5.2)$$

for Q. Now this central series can be written

$$1 = P_1/Z(P) \leq P_2/Z(P) \leq \ldots \leq P_n/Z(P) = P/Z(P) \qquad (5.3)$$

where each P_i is a normal subgroup of P and is such that $Z(P) \leq P_i \leq P$. The fact that the central series (5.3) is increasing forces

$$1 = P_0 \leq P_1 \leq P_2 \leq \ldots \leq P_n = P. \qquad (5.4)$$

Note how we have quietly smuggled in a definition of P_0, and have observed that $P_n/Z(P) = P/Z(P)$ forces $P_n = P$. Also notice that $P_1 = Z(P)$ and that the Correspondence Principle ensures that (5.4) is a normal series for P. The fact that (5.3) is a central series means that $[P_i x, P_i g] \in P_i$ for every $x \in P_{i+1}$ and every $g \in P$. However $[P_i x, P_i g] \in P_i$ is equivalent to $P_i[x, g] = P_i$ which in turn is the statement that $[x, g] \in P_i$. Since this holds for every $x \in P_{i+1}$ and for every $g \in G$ whenever $i \geq 1$, and $P_1 = Z(P)$, it follows that the series (5.4) is central and so P is nilpotent.

□

We now recall a result from the theory of finite p-groups. A proper subgroup of a finite p-group is strictly contained in its normalizer (Proposition 3.40). This is no idle observation, but was crucial in our proof of Sylow's theorem. Since nilpotent groups are in a sense generalizations of finite p-groups, it is perhaps not surprising that they inherit this important property.

Lemma 5.17

Let G be a nilpotent group and suppose that $H < G$ is a proper subgroup of H. It follows that the normalizer of H in G is strictly larger than H.

Proof

The result is true for abelian groups so we may induct on the nilpotency class of G and assume that G has class $c > 1$. Thus there is a central series

$$1 = G_0 \leq G_1 \leq G_2 \leq \ldots \leq G_c = G \tag{5.5}$$

If $G_1 \not\leq H$, then H is strictly contained in HG_1, and since elements of $H \cup G_1$ normalize H we deduce that $H \lhd HG_1$. Thus we are done unless $G_1 \leq H < G$. Let $\overline{G_i} = G_i/G_1$ for each $i \geq 1$ and put $\overline{H} = H/G_1$. Note that $\overline{G} = G/G_1$. The series

$$1 = \overline{G_1} \leq \overline{G_2} \leq \ldots \leq \overline{G_c} = \overline{G} \tag{5.6}$$

will be central by commutator arguments, so $\overline{G}$ has class $c-1$. Now $\overline{H}$ is a proper subgroup of $\overline{G}$ so by inductive hypothesis we have $\overline{H}$ is strictly contained in $N_{\overline{G}}(\overline{H})$. Now the Correspondence Principle yields that H is strictly contained in $N_G(H)$.

□

EXERCISES

5.7 Use the Correspondence Principle to prove Proposition 5.16 without using even a single commutator.

5.8 Find the composition factors for each finite symmetric group S_n.

5.9 Let G be a nilpotent group, and suppose that $H \leq G$. Prove that G has a subnormal series which has H as one of the groups in the series.

5.10 Let M be a maximal subgroup of a nilpotent group G. Prove that $G' \leq M$.

5.4 Varieties of Groups

The collection of all abelian groups are precisely those which obey the *law* $[x, y]$. In other words, G is abelian precisely when $[g, h] = 1$ for every $g, h \in G$. Similarly the collection of groups of exponent n consists of those groups which obey the law x^n. In general, a law is an expression which looks like a product in a group, but the letters of which it is composed are to be thought of as unknowns or variables. We test whether a particular law is satisfied by a group G by making all possible substitutions into the law from the group, and checking that each substitution yields the value 1.

Definition 5.18

Let L be a collection of laws. The *variety* $\mathfrak{V}_L$ of groups is the collection of all groups which obey every law in L.

Remark 5.19

The notation $\mathfrak{V}_L$ has just cropped up. It is a gothic V with an L subscript. We write $G \in \mathfrak{V}_L$ to mean that G is in the variety $\mathfrak{V}_L$ even though $\mathfrak{V}_L$ is not a set (there is a "set of all sets" problem here, there are even too many trivial groups for them safely to inhabit a set). Notice that if $G \in \mathfrak{V}_L$, then all subgroups and homomorphic images of G are in $\mathfrak{V}_L$. We say that $\mathfrak{V}_L$ is closed under forming subgroups and quotients. Moreover $\mathfrak{V}_L$ is also closed under the formation of Cartesian products (not just of pairs of groups, a variety is also closed under the formation of arbitrary Cartesian products, so we allow uncountably many factors).

In particular, it may be that $L = \{x^{-2}yx^2y, x^8\}$. The variety defined by these laws consists of those groups of exponent (dividing) 8 which have the property that squares of elements are central. Subgroups, homomorphic images and

Cartesian products of such groups of course obey the same laws.

Definition 5.20

Suppose that G is a group and that $g_1, g_2, \ldots, g_n \in G$. Define the *generalized* or *higher commutator* $[g_1, g_2, \ldots, g_n]$ inductively to be

$$[[g_1, g_2, \ldots, g_{n-1}], g_n] \text{ for } n \geq 3.$$

Proposition 5.21

A group G is nilpotent of class (at most) c if and only if it obeys the law $[x_1, x_2, \ldots, x_c, x_{c+1}]$.

Proof

Suppose that G is nilpotent of class c, so it has a central series

$$1 = G_0 \leq G_1 \leq G_2 \leq \ldots \leq G_c = G. \tag{5.7}$$

We prove by (finite) induction on n that $[g_0, g_1, \ldots, g_n] \in G_{c-n}$ for n in the range $1 \leq n \leq c$ and for all $g_i \in G$ where $0 \leq i \leq n$. The induction begins by observing that $[g_0, g_1] \in G_{c-1}$ for all $g_0, g_1 \in G$ since G/G_{c-1} is abelian. Now the inductive step is very similar; assuming that $[g_0, g_1, \ldots, g_m] \in G_{c-m}$ for all g_i, then the fact that (5.7) is a central ensures that $[g_0, g_1, \ldots, g_{m+1}] \in G_{c-m-1}$ for all g_i, and we are half way home.

Now instead suppose that G obeys the law $[x_1, x_2, \ldots, x_c, x_{c+1}]$. We put $\gamma_1(G) = G$ and for $i > 1$ put

$$\gamma_i(G) = \langle [g_1, g_2, \ldots, g_i] \mid g_a \in G \text{ for } 1 \leq a \leq i \rangle.$$

Note that if $\theta : G \to G$ is an endomorphism, then $([g_1, g_2, \ldots, g_i])\theta = ([(g_1)\theta, (g_2)\theta, \ldots, (g_i)\theta])$ so each $\gamma_i(G)$ is a fully invariant subgroup of G. In particular $\gamma_i(G) \trianglelefteq G$ for every i. Note that $\gamma_{c+1}(G) = 1$ since G obeys the law! Observe that for each $i \geq 2$ we have $[g_1, g_2, \ldots, g_i] \in \gamma_i(G) \trianglelefteq G$ for all $g_1, \ldots, g_i \in G$ and so

$$[[g_1, g_2, \ldots, g_i], g_{i+1}] = [g_1, g_2, \ldots, g_i]^{-1}[g_1, g_2, \ldots, g_i]^{g_{i+1}} \in \gamma_i(G).$$

Thus we have a normal series

$$1 = \gamma_{c+1}(G) \leq \gamma_c(G) \leq \cdots \leq \gamma_2(G) \leq \gamma_1(G) = G. \tag{5.8}$$

Now the statement that

$$[g_1, g_2, \ldots, g_i] = [[g_1, g_2, \ldots, g_{i-1}], g_i] \in \gamma_i(G)$$

for all values of the arguments simply asserts that

$$[\gamma_i(G)[g_1, g_2, \ldots, g_{i-1}], \gamma_i(G)g_i] = \gamma_i(G)$$

as $g_1, \ldots, g_i$ range over G. In other words, the image of our generating set for $\gamma_{i-1}(G)$ in $G/\gamma_i(G)$ is central, so that the normal series (5.8) is in fact a central series, and we are done.

$\square$

Corollary 5.22

The nilpotent groups of class c form a variety. Thus subgroups, factor groups and Cartesian products of nilpotent groups of class c are also nilpotent groups of class c.

Proposition 5.23

A finite group G is nilpotent if and only if it has exactly one Sylow p-subgroup for each prime number p dividing $|G|$.

Proof

Suppose that G is nilpotent. If G is the trivial group, then there are no prime numbers dividing $|G|$ and the result is vacuously true. Thus we may assume that G is not the trivial group and that there is a prime number p which divides $|G|$. Choose $P \in \mathrm{Syl}_p(G)$ and put $L = N_G(P)$. Suppose $g \in N_G(L)$, so that $L^g = L$. Thus conjugation by g induces an automorphism of L and $P^g \in \mathrm{Syl}_p(L)$. Thus there is $y \in L$ such that $P^g = P^y$ by Sylow's theorem. Now $P^{gy^{-1}} = P$ so $gy^{-1} \in N_G(P) = L$ so $g \in Ly = L$. Thus $N_G(L) \leq L$. However, it is a formality that $L \leq N_G(L)$ so $L = N_G(L)$. Since the group G is nilpotent, L is not a proper subgroup of G, so $N_G(P) = G$ and therefore $P \trianglelefteq G$. Thus P is the unique Sylow p-subgroup of G and we are half way home.

If G has just one Sylow p-subgroup for each prime number p dividing $|G|$, each of them is normal in G. The product of these Sylow p-subgroups is their internal direct product by an inductive argument based on the discussion surrounding Definition 2.36. However, this product is actually G by order considerations. Thus G is a Cartesian product of (finitely many) nilpotent groups of class at most c, and so G is nilpotent of class at most c.

Thus the finite nilpotent groups are the finite p-groups, and groups you can build from them using finite Cartesian products. Incidentally, the Cartesian product of infinitely many nilpotent groups is not nilpotent unless there is a bound on the classes of the factors.

We now investigate infinite nilpotent groups.

Definition 5.24

Let V be a group with normal series

$$1 = V_0 \leq V_1 \leq V_2 \leq \cdots \leq V_m = V. \tag{5.9}$$

Suppose that $\alpha \in \operatorname{Aut} V$ has the property that for every i in the range $1 \leq i \leq m$ and every $w \in V_i$ we have $V_{i-1}(w)\alpha = V_{i-1}w$. We say that α acts *nilpotently* on the given normal series for V.

In the notation of the definition, observe that each V_i is invariant under α (if $w \in V_i$, then $V_{i-1}(w)\alpha = V_{i-1}w$ so $(w)\alpha \in V_{i-1}w \subseteq V_i$). In the proof of the next proposition, we will use exponential notation for the application of an automorphism, so $(w)\alpha$ will be written as w^α. Note that $(w^{-1})^\alpha = (w^\alpha)^{-1}$ so we can write $w^{-\alpha}$ for either without risk of ambiguity. This notational change is not logically necessary, but raising the automorphisms above the line of print keeps them visually separated and from their arguments. We discussed this point at greater length in Remark 2.23 and the reader is urged to look back at that discussion.

Before we proceed, it is worth pointing out that there is another way to interpret a nilpotent action. Suppose that G is a group and that $K \trianglelefteq L \leq G$. Now suppose that $\theta \in \operatorname{Aut} G$ has the property that K and L are both θ-invariant ($K^\theta = K$ and $L^\theta = L$). It follows that θ induces an automorphism $\bar{\theta}$ of L/K defined by $Kx \mapsto K(x^\theta)$ for every $x \in L$. The enthusiastic reader should check the details. In this language, we can recast the statement that α acts nilpotently on the series (5.9) as follows: each term of the series is α invariant, and the induced automorphisms on the factors of the series are all identity maps.

Definition 5.25

If all elements of a group $A \leq \operatorname{Aut} V$ act nilpotently on the fixed normal series (5.9) of the group V, we say that A acts nilpotently on the normal series, and more casually, that A acts nilpotently on V.

Proposition 5.26

Let V be a group with normal series

$$1 = V_0 \leq V_1 \leq V_2 \leq \cdots \leq V_m = V. \tag{5.10}$$

Suppose that G is a group of automorphisms of V where G acts nilpotently on (5.10). It follows that G is a nilpotent group of class $m - 1$.

Proof

For each r in the range $1 \leq r \leq m - 1$ let G_r be the subset of G consisting of those elements of G which act nilpotently on every series obtained from (5.10) by omitting $r - 1$ consecutive groups V_i where $1 \leq i \leq m - 1$. Thus G_m is the trivial group and $G_1 = G$. It is straightforward to verify that each G_r is a subgroup of G, and we leave that to the reader. This condition about the action on edited versions of (5.10) amounts to the following; G_r consists of those elements of G which have the property that whenever $0 \leq i, i + r \leq m$ and $w \in V_{i+r}$, we have $wV_i = w^g V_i$. (Here we are using w^g instead of $(w)g$.) From this point of view it is clear that the conditions defining G_r become more strict as r increases so

$$1 = G_m \leq G_{m-1} \leq \cdots \leq G_1 = G \tag{5.11}$$

is a finite ascending sequence of subgroups of G. The balance of this proof will consist of showing that (5.11) is a central series for G. It will suffice to show that if $g \in G$ and $h \in G_r$, for $1 \leq r < m$ then $[h, g] \in G_{r-1}$. This will demonstrate both that each $G_j \trianglelefteq G$, and that the factors G_j/G_{j+1} are central in G/G_{j+1} for each j in the range $1 \leq j \leq m - 1$.

Choose any $w \in V_{i+r}$ where $i \geq 1$. We will show that $w^{hg}V_{i-1} = w^{gh}V_{i-1}$ for then replacing w by $w^{g^{-1}h^{-1}}$ it will follow that $w^{[g,h]}V_{i-1} = wV_{i-1}$ for every $w \in V_{i+r}$. This will ensure that $[g, h] \in G_{r+1}$ and the proof will be over.

Now the fun can begin. We will do some proper algebra. Note that $w^{-1}w^g \in V_{i+r-1}$ since g acts nilpotently on (5.10). Now $h \in G_r$ so

$$(w^{-1}w^g)^{-1}(w^{-1}w^g)^h \in V_{i-1}$$

and therefore in G/V_{i-1} we have

$$w^{gh}V_{i-1} = w^g w^{-1} w^h V_{i-1}. \tag{5.12}$$

Observe that $w^{-1}w^h \in V_i$ so $(w^{-1}w^h)^g V_{i-1} = w^{-1}w^h V_{i-1}$. Therefore equation (5.12) can become a plaything. We have

$$\begin{aligned} w^{gh}V_{i-1} &= w^g w^{-1} w^h V_{i-1} = w^g V_{i-1} \cdot w^{-1} w^h V_{i-1} \\ &= w^g V_{i-1} \cdot (w^{-1}w^h)^g V_{i-1} = w^{hg}V_{i-1}. \end{aligned}$$

$\square$

Example 5.27

Let V be an n dimensional vector space over $\mathbb{R}$. Select a basis $v_1, v_2, \ldots, v_n$. Define subspaces of V via $V_i = \text{span}\{v_1, v_2, \ldots, v_i\}$. Thus $\dim V_i = i$ and $0 = V_0 \le V_1 \le \cdots \le V_n = V$ is a normal series for V. Consider G the subset of $\text{GL}(V)$ consisting of the elements which act nilpotently on this series. It is straightforward to check that G is a group. By Proposition 5.26 G is a nilpotent group of nilpotency class $n - 1$. We can represent G as a group of matrices by picking a basis of V, but we did that already. If $g \in G$ then the matrix (g_{ij}) of g is obtained from the equations

$$gv_j = \sum_{i=1}^{n} g_{ij} v_i \text{ for } j = 1, \ldots, n.$$

The fact that g acts nilpotently forces $g_{ii} = 1$ for every i and $g_{ij} = 0$ if $i < j$. Thus G corresponds to the matrix group $\text{Tr}_1(n, \mathbb{R})$ which consists of the *upper unitriangular* matrices with shape

$$\begin{pmatrix}
1 & * & * & \cdot & \cdot & \cdot & * & * & * & * \\
0 & 1 & * & * & \cdot & \cdot & * & * & * & * \\
0 & 0 & 1 & * & * & \cdot & \cdot & * & * & * \\
0 & 0 & 0 & 1 & * & * & \cdot & \cdot & * & * \\
\cdot & \cdot & \cdot & \cdot & \cdot & * & \cdot & \cdot & \cdot & \cdot \\
\cdot & \cdot & \cdot & \cdot & \cdot & \cdot & * & \cdot & \cdot & \cdot \\
\cdot & \cdot & \cdot & \cdot & \cdot & 0 & 1 & * & * & * \\
\cdot & \cdot & \cdot & \cdot & \cdot & \cdot & 0 & 1 & * & * \\
0 & \cdot & \cdot & \cdot & \cdot & \cdot & \cdot & 0 & 1 & * \\
0 & 0 & \cdot & \cdot & \cdot & \cdot & \cdot & \cdot & 0 & 1
\end{pmatrix}$$

where the asterisks indicate that an entry may be any real number. Now the matrix equations yield that the matrices of elements of G_2 are exactly those with shape

$$\begin{pmatrix}
1 & 0 & * & \cdot & \cdot & \cdot & * & * & * & * \\
0 & 1 & 0 & * & \cdot & \cdot & * & * & * & * \\
0 & 0 & 1 & 0 & \cdot & \cdot & \cdot & * & * & * \\
0 & 0 & 0 & 1 & \cdot & \cdot & \cdot & \cdot & * & * \\
\cdot & \cdot & \cdot & \cdot & \cdot & \cdot & \cdot & \cdot & \cdot & \cdot \\
\cdot & \cdot & \cdot & \cdot & \cdot & \cdot & \cdot & \cdot & \cdot & \cdot \\
\cdot & \cdot & \cdot & \cdot & \cdot & 0 & 1 & 0 & * & * \\
0 & \cdot & \cdot & \cdot & \cdot & \cdot & 0 & 1 & 0 & * \\
0 & 0 & \cdot & \cdot & \cdot & \cdot & \cdot & 0 & 0 & 1 & 0 \\
0 & 0 & 0 & \cdot & \cdot & \cdot & \cdot & 0 & 0 & 0 & 1
\end{pmatrix}$$

where there is a superdiagonal of zeros. The matrices of G_3 have two super-diagonal rows of zeros and so on. We deduce that these sets of matrices from subgroups of $\mathrm{Tr}_1(n, \mathbb{R})$ and comprise a central series. We can vary $\mathbb{R}$ to any field, and obtain a similar central series for an upper unitriangular group of matrices. If we take V to be $\mathbb{Z}^n = \mathbb{Z} \oplus \mathbb{Z} \oplus \cdots \oplus \mathbb{Z}$ and use the "basis" $e_1, e_2, \ldots, e_n$ where $e_i = (0, 0, \ldots 0, 1, 0, \ldots 0)$ has a 1 in the i-th position and zeros elsewhere, then we obtain the nilpotent matrix group $\mathrm{Tr}_1(n, \mathbb{Z})$.

The remarkable Proposition 5.26 will have important applications later, but first we need to develop some more theory.

Proposition 5.28

Suppose that G is a nilpotent group and that G^{ab} is finite, then G is finite.

Proof

The sensible way to prove this is to use commutator arguments. We will give a proof which may be more amusing. If $G = 1$, then G is finite so we may assume that $G \neq 1$. Now G has class c and so has a central series of the form

$$1 = G_0 \leq G_1 \leq \ldots \leq G_c = G \qquad (5.13)$$

with $c \geq 1$. Now $(G/G_1)' = G'G_1/G_1$ so

$$(G/G_1)^{ab} = (G/G_1)/(G'G_1/G_1) \simeq G/G'G_1$$

but $G' \leq G'G_1$ so there is a natural epimorphism

$$G^{ab} = G/G' \to G/G'G_1 \simeq (G/G_1)^{ab}$$

by the third isomorphism theorem so $(G/G_1)^{ab}$ is finite. Now the result under discussion holds for G/G_1 by induction on class, and so G/G_1 is finite. Now Schur's Proposition 4.24 ensures that G' is finite, so $|G| = |G : G'||G'| < \infty$ and we are done.

$\square$

The next result is easy to prove when the group is abelian, but it is less obvious that it holds when the group is merely nilpotent. There must be some restriction on the group since the infinite dihedral group has an element of infinite order which is the product of a pair of involutions.

Proposition 5.29

Suppose that G is a nilpotent group. The elements of finite order form a subgroup of G.

Proof

The inverse of an element of finite order has finite order. It remains to show that if $x, y \in G$ have finite order, then xy has finite order. Let $H = \langle x, y \rangle$. Now H is a subgroup of G and so is nilpotent. The group H^{ab} is an abelian group generated by two elements of finite order, and so must be finite. We now deploy Proposition 5.28 so H is finite, but $xy \in H$ so xy is of finite order and we are done.

$\square$

5.5 Upper and Lower Central Series

In the proof of Proposition 5.21 we constructed a descending sequence of fully invariant subgroups (more or less) as follows. Consider the sequence of words (γ_i) where $\gamma_1 = x_1$ and for $i > 1$ we put $\gamma_i = [\gamma_{i-1}, x_i]$. These words are purely formal objects. Thus γ_i is a word in i formal variables and their symbolic inverses. Now suppose that G is a group, and for each i we define $\gamma_i(G)$ to be the group generated by all possible substitutions into the word γ_i from the group G. More formally

$$\gamma_i(G) = \langle [g_1, g_2, \ldots, g_i] \mid g_j \in G \text{ for } 1 \leq j \leq i \rangle. \tag{5.14}$$

We showed that each γ_i is a fully invariant subgroup of G, and that the sequence of groups descends in the sense that $\gamma_i(G) \leq \gamma_{i-1}(G)$ whenever $i > 1$, and moreover that $\gamma_{i-1}(G)/\gamma_i(G)$ is central in $G/\gamma_i(G)$. Thus this descending sequence of groups is trying very hard to be a central series for G. It can fail in two ways. It may be that the sequence descends forever, something which can and does happen in some infinite groups. Another possibility is that $\gamma_i(G) = \gamma_j(G) \neq 1$ for some $i \in \mathbb{N}$ and every integer $j \geq i$.

EXERCISES

5.11 Suppose that H is a subgroup of a group G. Prove that $\gamma_k(H) \leq \gamma_k(G)$ for every $k \in \mathbb{N}$.

5.12 Let $\theta : G \to H$ be a group homomorphism. Suppose that $g \in \gamma_k(G)$ for some $k \in \mathbb{N}$. Prove that $(g)\theta \in \gamma_k(H)$.

5.13 Suppose that N is a normal subgroup of a group G and that $k \in \mathbb{N}$. Prove that $\gamma_k(G/N) = \gamma_k(G)N/N$.

Proposition 5.30

Suppose that G is a group.

(i) Let $G = G_1 \unrhd G_2 \unrhd \cdots$ be an infinite descending sequence of normal
    subgroups of G with central factors. It follows that $\gamma_n(G) \leq G_n$ for every
    $n \in \mathbb{N}$.

(ii) G is nilpotent if and only if $\gamma_n(G) = 1$ for some $n \in \mathbb{N}$.

Proof

(i) We prove this by induction on n, the case $n = 1$ being given to us free by
    definition. Thus we may assume that $\gamma_{k-1}(G) \leq G_{k-1}$ for some $k > 1$. It
    follows that $[g_1, \ldots, g_{k-1}] \in G_{k-1}$ for all $g_1, \ldots, g_{k-1} \in G$. Now $[x, g] \in$
    G_k for all $x \in G_{k-1}$ and for all $g \in G$, so $[g_1, g_2, \ldots, g_k] \in G_k$ for all
    $g_1, \ldots, g_k \in G$. Therefore $\gamma_k(G) \leq G_k$.

(ii) If G is nilpotent, part (i) shows that $\gamma_n(G) = 1$ for some $n \in \mathbb{N}$. The
     proof of Proposition 5.21 shows that if $\gamma_n(G) = 1$ for some $n \in \mathbb{N}$, then
     G is nilpotent. Conversely if $\gamma_n(G) = 1$ for some $n \in \mathbb{N}$, then as we have
     previously established, the lower central series yields that G is nilpotent.

$\square$

Corollary 5.31

A group G is nilpotent of class c if and only if $\gamma_{c+1}(G) = 1$. Thus if G is
nilpotent, then the least value of c such that $\gamma_{c+1}(G) = 1$ is its its nilpotency
class (or pedantically, the least possible value of its nilpotency class).

The lower central series dives down as quickly as possible among all descending
sequences of normal subgroups of G with central factors. There is a dual notion
to this.

Definition 5.32

Let G be a group. Let $Z_0(G) = 1$ and for $n \in \mathbb{N}$ let $Z_n(G)$ be the
unique subgroup such that $Z_{n-1}(G) \leq Z_n(G) \leq G$ and $Z_n(G)/Z_{n-1}(G) = Z(G/Z_{n-1}(G))$.

Notice that the construction forces the infinite sequence to be ascending:

$$1 = Z_0(G) \leq Z_1(G) \leq Z_2(G) \ldots \tag{5.15}$$

and since each $Z_n(G)/Z_{n-1}(G) \trianglelefteq G/Z_{n-1}(G)$ it follows that each $Z_n(G) \trianglelefteq G$ by the Correspondence Principle. The series (5.15) is central by design. If ever $Z_n(G) = G$, then G must be nilpotent.

Proposition 5.33

Suppose that G is a group and that

$$1 = G_0 \leq G_1 \leq G_2 \leq \ldots$$

be an ascending sequence of normal subgroups of G with central factors. It follows that $G_n \leq Z_n = Z_n(G)$ for every $n \in \mathbb{N} \cup \{0\}$.

Proof

We proceed by induction on n, the result being true when $n = 0$. Thus we assume that $G_{k-1} \leq Z_{k-1}$ for some $k \in \mathbb{N}$. Now if $x \in G_k$, then $[x, g] \in G_{k-1} \leq Z_{k-1}$ for every $g \in G$, so $[Z_{k-1}x, Z_{k-1}g] = Z_{k-1}[x, g] = Z_{k-1}$ for every $g \in G$. Therefore $Z_{k-1}x \in Z(G/Z_{k-1}) = Z_k/Z_{k-1}$. Thus $x \in Z_k$ so $G_k \leq Z_k$ and we are done.

$\square$

Corollary 5.34

If G is nilpotent, then $Z_c(G) = G$ for some $c \in \mathbb{N} \cup \{0\}$. Moreover, if c is the least integer such that $Z_c(G)$ then G has nilpotency class c and is not nilpotent of class less than c.

Corollary 5.35

Suppose that G is nilpotent of class exactly c, and that

$$1 = G_0 < G_1 < \ldots < G_i < \ldots < G_c = G$$

is a central series manifesting this fact, then for each i in the range $0 \leq i \leq c$ we have $\gamma_{c+1-i}(G) \leq G_i \leq Z_i(G)$. This result flows from Corollary 5.34 and Proposition 5.30.

Definition 5.36

Suppose that G is a group and that $H, K \leq G$. We let

$$[H, K] = \langle [h, k] \mid h \in H, k \in K \rangle.$$

Thus $[G, G] = \gamma_2(G) = G'$.

Proposition 5.37

Suppose that G is a nilpotent group of class c. As usual let $(\gamma_i(G))$, $(Z_j(G))$ denote (respectively) the lower and upper central series of G.

(i) For all $i, j \in \mathbb{N}$ we have $[\gamma_i(G), \gamma_j(G)] \leq \gamma_{i+j}(G)$.

(ii) For all $i \in \mathbb{N}$ we have $[Z_i(G), \gamma_j(G)] \leq Z_{i-j}$ whenever $1 \leq j \leq i$.

Proof

Let $\tau : G \to \operatorname{Aut} G$ be the map $\tau : g \mapsto \tau_g$ where $(x)\tau_g = g^{-1}xg$ for every $x \in G$. Thus τ is a group homomorphism which we discussed in Section 2.5. Let $A = \operatorname{Inn} G = \operatorname{Im} \tau$. Now A acts nilpotently on any central series

$$1 = X_0 \leq X_1 \leq \cdots \leq X_m = G \tag{5.16}$$

for G. The proof of Proposition 5.26 gives a central series

$$1 = A_m \leq A_{m-1} \leq \cdots \leq A_1 = A \tag{5.17}$$

for A. Now the definition of A_j ensures that for all $x \in X_i$ and all $a \in A_j$ we have $x^{-1}x^a = (x^a x^{-1})^x \in X_{i-j}$, and this is true for all subscripts i and j with the indulgence that negative subscripts are read as 0. Now let G_j be the τ-preimage of A_j. For $x \in X_i$ and $g \in G_j$ we have

$$x^{-1}x^{\tau_g} = [x, g] \in X_{i-j}.$$

Also note that

$$1 \leq Z(G) = G_m \leq G_{m-1} \leq \cdots \leq G_1 = G \tag{5.18}$$

is a central series for G.

Now we can quickly finish both parts of the question. For part (i) we let (5.16) be the lower central series of G. In this case $m = c$ and we deduce that $[\gamma_i(G), G_j] \leq \gamma_{i+j}(G)$ for all i and j. However, since the series (5.18) is central, $\gamma_j(G) \leq G_j$ and therefore $[\gamma_i(G), \gamma_j(G)] \leq \gamma_{i+j}(G)$ for all integers i and j.

As for part (ii), we let (5.16) be the upper central series of G, and argue in exactly the same way.

$\square$

We observed earlier that if G is nilpotent and G^{ab} is finite then G is finite. Thus G^{ab} infects G with finiteness. We will now see that the centre of a nilpotent group is also generous in propagating its properties.

Proposition 5.38

Suppose that G is a nilpotent group of class c and that $Z(G)$ has finite exponent m. It follows that G has exponent (dividing) m^c.

Proof

We work with the upper central series (Z_i) of G. Choose any term Z_{i+1} of this series with $i \geq 1$ and suppose that $g \in G$. Define a map $\psi_g : Z_{i+1} \to Z_i/Z_{i-1}$ by $t \mapsto [t,g]Z_{i-1}$ for every $t \in Z_{i+1}$. By Philip Hall's formula

$$(xy)\psi_g = [xy,g]Z_{i-1} = [x,g]^y[y,g]Z_{i-1} = [x,g][y,g]Z_{i-1}.$$

The last equality holds because $[x,g]Z_{i-1} \in Z_i/Z_{i-1}$ and so is central in G/Z_{i-1}. Thus $\psi_g : Z_{i+1} \to Z_i/Z_{i-1}$ is a group homomorphism. Notice that $Z_i \subseteq \text{Ker } \psi_g$ so by the third isomorphism theorem (Theorem 2.29) we have an induced natural homomorphism $\widehat{\psi}_g : Z_{i+1}/Z_i \to Z_i/Z_{i-1}$ defined by $tZ_i \mapsto [t,g]Z_{i-1}$.

Having set up this little machine, we are ready for the proof. We show that each factor Z_n/Z_{n-1} has exponent m by induction on n, the case $n = 1$ being given. Now suppose that the result holds when $n = i$, and address the case $i + 1$. For each $g \in G$ we have the homomorphism $\psi_g : Z_{i+1} \to Z_i/Z_{i-1}$ defined above. Suppose that $x \in Z_{i+1}$ so $(x^m)\psi_g = (x)\psi_g^m = 1$ since Z_i/Z_{i-1} has exponent m. Therefore $[x^m,g] \in Z_{i-1}$ for every $g \in G$ so $x^m Z_{i-1} \in Z(G/Z_{i-1})$. Therefore $x^m \in Z_i$. However, the choice of $x \in Z_{i+1}$ was arbitrary so Z_{i+1}/Z_i has exponent m. By finite induction each factor of the upper central series has exponent m. If $u \in G$ then $u^{m^j} \in Z_{c-j}$ by induction on j, and so G has finite exponent (dividing) m^c.

$\square$

Proposition 5.39

Let $G = \langle A \rangle$ be a nilpotent group with lower central series $(\gamma_i(G))$. Suppose that $\gamma_{i-1}(G) = \langle B \rangle \trianglelefteq G$ where the generating sets A and B are both closed under inversion.

(i) Let $R = \langle [h,g] \mid h \in \gamma_{i-1}(G), g \in G \rangle$. It follows that $R = \gamma_i(G)$.

(ii) The group $\gamma_i(G)/\gamma_{i+1}(G)$ is generated by $\{[b,a]\gamma_{i+1}(G) \mid b \in B, a \in A\}$.

Proof

(i) Since $\gamma_{i-1}(G) \trianglelefteq G$ it follows that $R \leq \gamma_{i-1}(G)$. Recall that we defined $\gamma_i(G)$ to be the group generated by all elements of the form

$$[g_1, g_2, \ldots, g_i] = [[g_1, \ldots, g_{i-1}], g_i]$$

for all arguments g_j. Each of these generators of $\gamma_i(G)$ is an element of the specified generating set for R, so $\gamma_i(G) \leq R \leq \gamma_{i-1}(G)$. However, for each $h \in \gamma_{i-1}(G)$ and $g \in G$ we have

$$[h, g]\gamma_i(G) = [h\gamma_i(G), g\gamma_i(G)] = \gamma_i(G)$$

since $\gamma_{i-1}(G)/\gamma_i(G)$ is central in $G/\gamma_i(G)$. We conclude that $R = \gamma_i(G)$.

(ii) Since A and B are closed under inversion, elements of G and $\gamma_{i-1}(G)$ can be written as words on A and B (respectively) without the need for negative exponents. Now for $x, y \in \gamma_{i-1}(G)$ and $z \in G$ we have

$$[xy, z]\gamma_{i+1}(G) = [x, z][x, z, y][y, z]\gamma_{i+1}(G) = [x, z][y, z]\gamma_{i+1}(G). \quad (5.19)$$

See the result of Exercise 5.1, p.160. The second equality in (5.19) holds because $[x, z] \in \gamma_i(G)$ so $[x, z, y] \in \gamma_{i+1}(G)$, and therefore

$$\begin{aligned}[x, z][x, z, y][y, z]\gamma_{i+1}(G) &= [x, z][y, z][x, z, y]^{[y, z]}\gamma_{i+1}(G) \\ &= [x, z][y, z]\gamma_{i+1}(G).\end{aligned}$$

Similarly Proposition 5.1 yields the identity $[z, xy] = [z, y][z, x][z, x, y]$ so if $z \in \gamma_{i-1}(G)$ and $x, y \in G$ we have

$$[z, xy]\gamma_{i+1}(G) = [z, y][z, x][z, x, y]\gamma_{i+1}(G) = [z, y][z, x]\gamma_{i+1}(G) \quad (5.20)$$

since $[z, x, y] \in \gamma_{i+1}(G)$. By part (i) we know that $\gamma_i(G)/\gamma_{i+1}(G)$ is generated by all the elements of the form $[h, g]$ with h a word (with non-negative exponents) on the elements of B and g a word (with non-negative exponents) on the elements of A. Now Eq.(5.19) and Eq.(5.20) together ensure that $\gamma_i(G)/\gamma_{i+1}(G)$ is generated by the elements of the form $[b, a]\gamma_{i+1}(G)$.

$\square$

Corollary 5.40

If G is a finitely generated nilpotent group, then $\gamma_i(G)/\gamma_{i+1}(G)$ is a finitely generated abelian group for every i.

Ths corollary follows by finite induction on i.

We will very soon be able to prove the beautiful result that a subgroup of a finitely generated nilpotent group is finitely generated.

EXERCISES

5.14 Suppose that G is a group and that for some integer k we have $\gamma_k(G) = \gamma_{k+1}(G)$. Prove that $\gamma_j(G) = \gamma_k(G)$ for every integer $j \geq k$.

5.15 Suppose that $H, K \trianglelefteq G$. Prove that $[H, K] \trianglelefteq G$ and that $[H, K] \leq H \cap K$.

5.6 Soluble Groups

Recall that a group G is soluble (solvable) if and only if it is polyabelian. We mentioned the class of polycyclic groups at the start of this chapter. Since cyclic groups are the most straightforward abelian groups there is reason to hope that the study of polycyclic groups will prove rewarding.

Proposition 5.41

If G is a finitely generated abelian group, then G is polycyclic.

Proof

This can be read off from the structure theorem, but it is easy enough to give a direct proof. Suppose that $\langle X \rangle = G$ and that X has minimum size among all generating sets for G. We induct on $|X|$. If $X = \emptyset$, then G is the trivial group and we are done. Thus we may assume that $X \neq \emptyset$ and select $x \in X$. The natural projection $G \to H = G/\langle x \rangle$ sends X to a generating set for H, but $x \mapsto 1$ and so may be omitted from this generating set for H and therefore H requires fewer than $|X|$ generators. Thus H is polycyclic. Suppose that

$$1 = H_1 \leq H_1 \leq \cdots \leq H_m = H$$

is a subnormal series for H with cyclic factors. By the correspondence principle each $H_i = G_i/\langle x \rangle$ with $\langle x \rangle \leq G_i \leq G$. Moreover $H_i \trianglelefteq H_{i+1}$ entails $G_i \trianglelefteq G_{i+1}$, and the third isomorphism theorem (Theorem 2.29) yields that each $G_{i+1}/G_i \simeq H_{i+1}/H_i$ is cyclic for i in the range $1 \leq i \leq m$. Now

$$1 = G_0 \leq G_1 = \langle x \rangle \leq G_2 \leq \cdots \leq G_m = G$$

is a subnormal series for G with cyclic factors, so G is polycyclic.

$\square$

Proposition 5.42

Let G be a polycyclic group.

(i) If $N \trianglelefteq G$ then G/N is polycyclic.

(ii) If $H \leq G$ then H is polycyclic.

Proof

Suppose that $1 = G_0 \trianglelefteq G_1 \trianglelefteq \cdots \trianglelefteq G_n = G$ is a subnormal series for G with cyclic factors.

(i) Suppose that $x \in G_i, y \in G_{i+1}$ and $m, n \in N$, then

$$(xm)^{yn} = (x^y)^n m^{yn} = x^y[x^y, n]m^{yn} = x[x,y][x^y, n]m^{yn} \in G_i N.$$

It follows that $G_i N \trianglelefteq G_{i+1} N$ for every i in the range $0 \leq i < n$. We have the inclusion map $G_{i+1} \to G_{i+1}N$ and the natural projection $G_{i+1}N \to G_{i+1}N/G_iN$. The composite of these two homomorphisms is a homomorphism $G_{i+1} \to G_{i+1}N/G_iN$. which is surjective by inspection, and G_i is contained in the kernel. Therefore by the third isomorphism theorem (Theorem 2.32) we have an induced natural epimorphism $G_{i+1}/G_i \to G_{i+1}N/G_iN$ which incidentally is just $xG_i \mapsto xG_iN$. Now G_{i+1}/G_i is cyclic, and the homomorphic image of a cyclic group is cyclic by Proposition 1.43, so $G_{i+1}N/G_iN$ is cyclic. The series

$$1 = G_0N/N \trianglelefteq G_1N/N \trianglelefteq \cdots \trianglelefteq G_nN/N = G/N$$

is a subnormal series for G/N and we have

$$(G_{i+1}N/N)/(G_iN/N) \simeq G_{i+1}N/G_iN$$

is cyclic by another application of the third isomorphism theorem (Theorem 2.29). Therefore G/N is polycyclic.

(ii) Let $H_i = G_i \cap H$ for every i. Now consider the composite θ of inclusion $H_{i+1} \to G_{i+1}$ and the natural epimorphism $G_{i+1} \to G_{i+1}/G_i$. Now θ has H_i as kernel and $H_{i+1}G_i/G_i$ as image. Kernels are normal in their domains, so $H_i \trianglelefteq H_{i+1}$ and the first isomorphism theorem tells us that $H_{i+1}/H_i \simeq H_{i+1}G_i/G_i$. However $H_{i+1}G_i/G_i \leq G_{i+1}/G_i$ which is cyclic. Now a subgroup of a cyclic group is cyclic by Proposition 1.43, so H_{i+1}/H_i is cyclic for every i and $1 = H_0 \trianglelefteq H_1 \trianglelefteq \cdots \trianglelefteq H_n = H$ is a subnormal series for H which demonstrates that H is polycyclic.

$\square$

Corollary 5.43

A subgroup of a polycylic group is polycyclic and therefore finitely generated. In particular, any subgroup of a finitely generated abelian group is a finitely generated abelian group.

Proposition 5.44

Suppose that G is a finitely generated nilpotent group, then G is polycyclic. H is a subgroup of G. It follows that H is finitely generated.

Proof

By Corollary 5.40 each factor $\gamma_i(G)/\gamma_{i+1}(G)$ is a finitely generated abelian group. Such groups are polycyclic by Proposition 5.41. Thus we can refine the lower central series of G by inserting extra groups between each $\gamma_{i+1}(G)$ and $\gamma_i(G)$ to produce a subnormal (normal in fact) series with cyclic factors. Therefore G is polycyclic and Corollary 5.43 sees us home.

□

Corollary 5.45

Any subgroup of a finitely generated nilpotent group is a finitely generated nilpotent group.

The minimal length of a subnormal series of G with abelian factors is called its soluble length. Groups of soluble length at most 2 are called *metabelian*. This piece of notation is decidedly odd, but somehow metabelian groups crop up a lot, so it does make sense to have a special name for them.

There seems to be no satisfactory analogue of the upper central series for soluble groups. The point is that the product of normal abelian subgroups is not necessarily abelian, whereas the intersection of normal subgroups with abelian factors is a normal subgroup with an abelian factor. There is a descending series which screams out for attention.

Define formal words as follows. Put $\delta_0 = x_1$. Now suppose that δ_i has been defined and $\delta_i(x_1, \ldots, x_{2^i})$ is a word in 2^i variables. Let

$$\delta_{i+1} = [\delta_i(x_1, \ldots, x_{2^i}), \delta_i(x_{2^i+1}, \ldots, x_{2^{i+1}})]$$

in 2^{i+1} variables. This is a complicated way of saying that

$$\delta_1 = [x_1, x_2],$$

$$\delta_2 = [[x_1, x_2], [x_3, x_4]],$$

$$\delta_3 = [[[x_1, x_2], [x_3, x_4]], [[x_5, x_6], [x_7, x_8]]]$$

and so on. We define $\delta_i(G)$ to be the subgroup of G generated by all possible substitutions into the word δ_i. It is an easy induction to show that $\delta_{i+1}(G)$ is a normal subgroup of $\delta_i(G)$. Moreover, the generators of $\delta_i(G)$ commute modulo $\delta_{i+1}(G)$ so $\delta_i(G)/\delta_{i+1}(G)$ is abelian. Moreover the manner of definition of these groups $\delta_i(G)$ ensures that they are fully invariant in G, and are therefore certainly normal in G.

Definition 5.46

Suppose that G is a group. The descending sequence of normal subgroups

$$G = \delta_0(G) \geq \delta_1(G) \geq \cdots$$

with abelian factors is called the derived series of G.

The definition of the derived series ensures that $G/\delta_j(G)$ is soluble of soluble length j. This is because each $\delta_j(G)$ is fully invariant (and so normal) in G, and the given generators of $\delta_j(G)$ commute in the factor group $\delta_j(G)/\delta_{j+1}(G)$.

Proposition 5.47

Suppose that G is a soluble group with a subnormal series

$$1 = G_n \leq G_{n-1} \leq \cdots \leq G_0 = G$$

with abelian factors. It follows that $\delta_i(G) \leq G_i$ for every i.

Proof

Induct on n.

□

Corollary 5.48

A soluble group is defined to be a group with a subnormal series which has abelian factors, but it now follows that such a group has a normal series (even a fully invariant series) with abelian factors. Moreover, the collection of soluble groups of soluble length d is a variety since it is defined by a law. Thus soluble groups of soluble length d are closed under forming subgroups, factor groups and Cartesian products.

Note that the derived series dives down into G faster than any other subnormal series with abelian factors.

EXERCISES

5.16 Let G be a group. Define a sequence of subgroups by $G_1 = G$, and for $i > 1$ put $G_i = [G_{i-1}, G_{i-1}]$ (in the sense of Definition 5.36). Prove that this is another way to construct the derived series.

5.17 Let G be a non-trivial finite soluble group. Suppose that N is a minimal normal subgroup of G (so if $H \trianglelefteq G$ and $1 < H \leq N$, then $H = N$). Show that N is the Cartesian product of finitely many copies of the group C_p where p is a prime number.

5.18 Let G be a non-trivial finite soluble group. Suppose that M is a maximal subgroup of G (so if $H < G$ and $M \leq H < G$, then $M = H$). Prove that $|G : M|$ has prime power order.

6
Presentations

6.1 Informalities

We start the theory of presentations with a relaxed chat. Suppose that we are told that a group G has generators a, b, c and we are also given the information that $abc = a^2 = b^2 = c^2 = 1$. Let us see what we can find out about G.

We have $ab = c^{-1}$ so $(ab)^2 = (c^2)^{-1} = 1$. Now $abab = 1$. Premultiply by a and postmultiply by b to discover that $aababb = ab$. However, $a^2 = b^2 = 1$ so $ba = ab$. Thus $b \in C_G(a)$, and moreover $c = (ab)^{-1} \in C_G(a)$. We conclude that $C_G(a) = G$ so $a \in Z(G)$. Similarly $b \in Z(G)$ so $c = (ab)^{-1} \in Z(G)$. It follows that

$$G = \langle a, b, c \rangle \leq Z(G) \leq G$$

so G is abelian.

Next we will show that G has at most 4 elements. We can eliminate c and c^{-1} by replacing any occurrence of c by $b^{-1}a^{-1}$ (which is $(ab)^{-1}$). Any word (product) in a, b, c and their inverses is equal to a word on just a and b since $a^{-1} = a$ and $b^{-1} = b$. We now use the fact that G is abelian to move all occurrences of a to the left of the word.

Thus any element of G is expressible as $a^m b^n$ where m, n are non-negative integers. We may replace m and n by their remainders on division by 2 since $a^2 = b^2 = 1$. Thus $G = \{a^0 b^0, a^1 b^0, a^0 b^1, a^1 b^1\} = \{1, a, b, ab\}$. We conclude that $|G| \leq 4$. We cannot conclude that $|G| = 4$ because we do not know that the elements $1, a, b$ and ab are distinct.

We are in a rather peculiar position. We know names for every element of G, but some of the names may in fact refer to the same group element. We also know how to multiply in G, since we have enough information to express the product of two elements drawn from $\{1, a, b, ab\}$ as an element of $\{1, a, b, ab\}$. We can write down a sort of Cayley table, where all the information in the table is correct, but some rows and columns may be redundant.

Here are three non-isomorphic groups which are all perfectly good candidates for G.

(i) $G \leq S_4$ and $a = (1,2)(3,4)$, $b = (1,3)(2,4)$, $c = ab = (1,4)(2,3)$. Here $G \simeq C_2 \times C_2$ is the Vierergruppe.

(ii) $G = S_2 \simeq C_2$ and $a = b = (1,2)$, $c = \mathrm{id}$.

(iii) G is the trivial group and $a = b = c = ab = 1$.

Now the important group in this list is given by (i). The existence of this group shows that the tentative Cayley table that we discussed for $\{1, a, b, ab\}$ does define an associative operation, and no collapse is forced.

Proposition 6.1

The three groups listed above are (up to isomorphism) the only groups G which are generated by elements a, b, c satisfying $abc = 1$ and $a^2 = b^2 = c^2 = 1$.

Proof

We may conjugate $abc = 1$ by a to deduce that $bca = 1$. We conjugate this in turn by b to obtain that $cab = 1$. Thus the data has more symmetry than first appears. Each of a, b and c has equal status.

We know that $G = \{1, a, b, ab\}$. Suppose that $c = 1$, then $G = \{1, b\}$ and $a = b$ (if instead of putting $c = 1$ we put $a = 1$ or $b = 1$ the situation is essentially the same because of the symmetry in the data specifying G). Now there are two possibilities: $a \neq 1$ and $a = 1$. These possibilities respectively yield the cyclic group of order 2 generated by a ($= b$) and the trivial group. We have accounted for the three examples on our list. We now examine other forms of collision in the knowledge that if at least one of a, b or c is 1 we have the situation under control.

There are three other forms of collapse: $a = b$, $b = ab$ and $a = ab$. However, if $a = b$ then $c = ab = a^2 = 1$. If $b = ab$, then $a = 1$. Finally if $a = ab$ then $b = 1$. Thus if $\{1, a, b, c\}$ does not have four elements then at least one of a, b, c is 1 and we are done.

$\square$

Remark 6.2

We used very elementary methods to prove Proposition 6.1, but there is a serene view from a nearby hill. Any group satisfying the given conditions will be a homomorphic image of the group discussed in (i) because the Cayley table will be just right. Thus the groups generated by elements a, b, c satisfying $abc = 1$ and $a^2 = b^2 = c^2 = 1$ are the factor groups of $C_2 \times C_2$. This leads to the three alternatives we found by direct calculation.

Remark 6.3

You feel that the group $C_2 \times C_2$ is somehow the "correct" group specified by the data. The other two groups involve equalities among the elements $1, a, b, c$ which cannot be deduced from the given equations. Why can these equalities not be deduced? Well, simply because the equalities do not hold in group (i). If additionally we specify that $c = 1$ we know that we are in the situation specified by either (ii) or (iii). Again, (ii) is the "correct" answer, because (iii) involved the extra data that $a = 1$. This information cannot be deduced from the data specifying (ii) since $a = 1$ is not true in the group specified by (ii).

In order to define the group described in (i) we could say:

The group G has generators a, b, c and you are also given the information that $abc = 1$ and $a^2 = b^2 = c^2 = 1$. Moreover, any equation in the group G is deducible from these equations and the group axioms.

When we have set up the theory properly, we will write the italicized sentences in the following economical way:

$$G = \langle a, b, c \mid a^2, b^2, c^2, abc \rangle.$$

We now move on to a superficially easier question. We specify a group F as follows:

The group F has generators x, y and you are also given no extra information. Moreover, any equation in the group G is deducible from the group axioms.

Elements of F are words in x, x^{-1}, y, y^{-1} with the usual conventions about g^n for $n \in \mathbb{Z}$ and $g \in F$. Thus

$$x^3 x^{-2} xy^{-7} xx^{-1} y^{10} y^{-6} x^{-1}$$

is an element of F. However, we can use the group axioms to eliminate juxtapositions of generators and their inverses (a process that we will call *free reduction*). Thus we discover that

$$x^3 x^{-2} xy^{-7} xx^{-1} y^{10} y^{-6} x^{-1} = x^2 y^{-3} x^{-1}$$

which you can verify in your head. Now there is a little problem here. There are various ways in which the free reduction can be performed on the left hand side. For example, we might begin by cancelling xx^{-1} or alternatively we might start by replacing $y^{10}y^{-6}$ by y^9y^{-5}. Fortunately it does not matter. No matter how we reduce the left side (until we get stuck), the procedure always terminates with the word $x^2y^{-3}x^{-1}$. We have not yet proved this. A word such as $x^2y^{-3}x^{-1}$ which admits of no further obvious cancellation is called a *freely reduced word*. We will save the formal definitions for later. In general an arbitrary word cancels to a unique freely reduced word.

Now suppose that instead of just cancelling, we are allowed to insert extra occurrences of juxtaposed generators with their inverses, so we can manipulate

$$x^3x^{-2}xy^{-7}xx^{-1}y^{10}y^{-6}x^{-1}$$

by replacing it by

$$x^3x^{-2}xy^{-7}x(y^{-1}y)x^{-1}y^{10}y^{-6}x^{-1}$$

and in the middle of the newly inserted pair we may insert another $y^{-1}y$ to obtain

$$x^3x^{-2}xy^{-7}x(y^{-2}y^2)x^{-1}y^{10}y^{-6}x^{-1}$$

and carry on in the same vein to obtain

$$x^3x^{-2}xy^{-7}x(y^{-42}y^{42})x^{-1}y^{10}y^{-6}x^{-1}.$$

There is nothing to stop us picking another spot and doing some insertions or cancellations there. This process seems very chaotic, and we can blunder around indefinitely creating group elements (words) which are all equal to the group element (word) that we started with. At any stage we have the option of forswearing the use of insertions, and cancelling in any order that we please. It may not be obvious, but in fact we still always get stuck at the reduced word $x^2y^{-3}x^{-1}$. There is nothing special about the word

$$x^3x^{-2}xy^{-7}xx^{-1}y^{10}y^{-6}x^{-1}$$

that we have been studying. In general the free reduction (cancellation) and insertion process applied to any word has the property that each word can be shown to be equal to a unique reduced word. Moreover, the fact that our generating set consists of two symbols is also irrelevant, and everything works just the same for an arbitrary set of generating symbols. All this is genial bullying (what passes for explanation in some quarters). We don't actually know what the group F is, though we have been happily manipulating its elements with touching faith.

So how shall we build F? One approach is to let F be the set of reduced words with multiplication defined by juxtaposition and reduction. Thus to multiply xyx^{-1} by $xy^{-1}x^{22}$ we write the words next to one another and then freely reduce. Thus

$$xyx^{-1} * xy^{-1}x^{22} = x^{23}.$$

It is easy to see that the empty word (written 1) is a two-sided identity element, and that the inverse of a (reduced) word is obtained by writing it backwards and changing the signs of the exponents. Associativity is awkward though, unless you have proved the theorem which says that reduction puts each group elements into a unique reduced form. Logically there is nothing wrong with this way of building F, but we think another way is slightly better.

In order to illustrate the difficulty in a familiar context, we will discuss how the rational numbers can be constructed.

6.2 The Rational Numbers

We take the integers $\mathbb{Z}$ as given. We first construct the rationals using the "reduced word" method outlined above.

6.2.1 The Rationals Mark I

Given any $a, b \in \mathbb{Z}$ with $b \neq 0$ we say that the formal sequence of symbols a/b (a slash b) is *reduced* if the greatest common divisor of a and b is 1. Note that we are NOT thinking of a/b as "a divided by b". At this stage it is simply a sequence of three symbols and nothing more.

If d is the greatest common divisor of a and b, then put $c = (bd) \div |b|$ so $c = \pm d$. We have arranged that c has the same sign as b. Now x/y is reduced where $xc = a$ and $yc = b$. The process of passing from a/b to x/y will be called *reduction* and we write $x/y = \mathrm{red}(a/b)$. Note that $y > 0$.

Let $\mathbb{Q}$ be the set of reduced expressions x/y. We define addition and multiplication in $\mathbb{Q}$ as follows. If $x_1/y_1, x_2/y_2 \in \mathbb{Q}$, then

$$x_1/y_1 + x_2/y_2 = \mathrm{red}((x_1y_2 + x_2y_1)/y_1y_2)$$

and

$$x_1/y_1 \times x_2/y_2 = \mathrm{red}(x_1x_2/y_1y_2).$$

The devoted student should then settle down with a mug of cocoa and verify that the object that we have just built is a field. The subset of $\mathbb{Q}$ consisting of elements of the form $x/1$ is a new copy of $\mathbb{Z}$ inside $\mathbb{Q}$. Moreover, each rational number x/y has the property that $x/1 = x/y \times y/1$. Thus if we discard our old integers and work with new integers instead, we can reactivate those neural pathways which want to think of a/b as a divided by b. The slightly sour aspect of this construction is that $2/4$ is not a rational number. Now 2 divided by 4 is $1/2$ because $4/1 \times 1/2 = \text{red}(4/2) = 2/1$. Also note that $1/2 = \text{red}(2/4)$.

From the point of view of mathematical elegance, this is a little unpleasant. We face the disappointment that $2/4$ is not a rational number, and the fact that $1/2 = \text{red}(2/4)$ is little consolation.

6.2.2 The Rationals Mark II

Now we begin again, and construct $\mathbb{Q}$ from $\mathbb{Z}$ a second time. This time we will use an equivalence relation. Let

$$\Omega = \{(a, b) \mid a, b \in \mathbb{Z}, b \neq 0\}.$$

Define a relation $\sim$ on Ω by $(a, b) \sim (c, d)$ if and only if $ad - bc = 0$. Honesty dictates that you check that this is an equivalence relation. Denote the equivalence class of (a, b) by the symbol a/b. Thus

$$1/2 = 2/4 = -13/26 = 22/(-44)$$

since these are all descriptions of the same equivalence class.

Addition and multiplication are defined as follows. If $x_1/y_1, x_2/y_2 \in \mathbb{Q}$, then

$$x_1/y_1 + x_2/y_2 = (x_1 y_2 + x_2 y_1)/y_1 y_2$$

and

$$x_1/y_1 \times x_2/y_2 = x_1 x_2/y_1 y_2$$

and the reduction process has disappeared. There is a price to be paid however. We must check that these operations are well defined. We (i.e. you) must verify that if $x_1/y_1 = x_1'/y_1'$ and $x_2/y_2 = x_2'/y_2'$, then

$$((x_1 y_2 + x_2 y_1), y_1 y_2) \sim ((x_1' y_2' + x_2' y_1'), y_1' y_2')$$

and

$$x_1 x_2/y_1 y_2 \sim x_1' x_2'/y_1' y_2'.$$

This will ensure that the answer we get when adding or multiplying x_1/y_1 and x_2/y_2 depends only on the equivalence class, and not on the particular way in which it is being described. When this is done, it is then possible to check that $\mathbb{Q}$ is a field with all the right properties, and that the rational numbers of the form $x/1$ are a copy of the integers.

This also illustrates how easy equivalence classes are. Surely the reader has been happy to write $6/4 = 3/2$ for most of his or her life. This is correct from our current perspective because $6/4$ and $3/2$ are rival descriptions for the same equivalence class.

6.3 Tigers

So, enough of the propaganda. We hope that you are persuaded that the mathematically cleaner approach is via equivalence relations.

We will construct the group F as a collection of equivalence classes. Instead of working with a two-letter alphabet $\{a, b\}$, we will use an arbitrary alphabet. Let Σ be a set of atoms. The unusual stipulation of atomicity is because we wish to consider Σ^*, the set of all words (i.e. sequences) of finite length on the alphabet Σ, and we do not want to confront the distressing situation that an element of Σ is actually a word on some other letters of Σ. Such unpleasantness can arise very easily if we do not exclude it, as in $\{a, aa\}$. Since the elements of Σ are atoms, each word $w = \sigma_1 \sigma_2 \ldots \sigma_n \in \Sigma^*$ is uniquely expressible as a string of elements $\sigma_i \in \Sigma$. From now on, the atomic proviso will always be implicit (i.e. we will not mention it). Juxtaposition of words is a closed binary operation on Σ^*. The algebraic structure we get is called a monoid and the empty word ε is a two sided identity.

The Monoid Axioms. A monoid $(M, *)$ consists of a set M, and a closed binary operation $*$ such that these axioms hold:

(i) (Associative law) For every $x, y, z \in M$ we have $(x * y) * z = x * (y * z)$.

(ii) (Identity) There is $e \in M$ such that for all $x \in M$ we have $e * x = x * e = x$.

A monoid is like a group in all respects save that there is no guarantee that inverses exist. Thus all groups are monoids, but not all monoids are groups. For example, the integers form a monoid under multiplication. We afforded considerable space in Chapter 1 to the proposition that the natural setting for group theory is the study of collections of bijections from a set to itself. When discussing monoid, it may be helpful to think of the set of all maps from a set to itself. This will form a monoid under composition of functions. The monoid counterpart of the symmetric group S_n therefore has size n^n.

Definition 6.4

Suppose that M and N are monoids. A map $\varphi : M \to N$ is a *monoid homomorphism* if the following two conditions are satisfied.

(a) For all $x, y \in M$ we have $(x)\varphi * (y)\varphi$.

(b) $(1_M)\varphi = 1_N$.

The group theoretic notions of monomorphism, epimorphism and isomorphism have their obvious counterparts in the theory of monoids. We will now drop $*$ and will use juxtaposition to denote a product in a monoid.

We now return to the discussion of the specific monoid Σ^*. We will discuss certain equivalence relations on this monoid, and the reader may wish to seek support from Appendix B.

Definition 6.5

An equivalence relation $\sim$ on Σ^* is *translation invariant* if whenever ω_1, $\omega_2, \lambda, \rho \in \Sigma^*$ and $\omega_1 \sim \omega_2$, then $\lambda\omega_1\rho \sim \lambda\omega_2\rho$.

If $\sim$ is a translation invariant equivalence relation, then the set of equivalence classes $\Sigma^* / \sim$ has an induced monoid structure via $[\omega_1][\omega_2] = [\omega_1\omega_2]$, but we have to check that this operation is well defined. Suppose that $\omega_1 \sim \omega_1'$ and $\omega_2 \sim \omega_2'$. Now we use translation invariance to see that $\omega_1\omega_2 \sim \omega_1\omega_2' \sim \omega_1'\omega_2'$ so all is well.

Remark 6.6

Suppose that $\sim_\alpha$ and $\sim_\beta$ are both translation invariant equivalence relations on Σ^* and that $\sim_\alpha \subseteq \sim_\beta$, so that the equivalence classes of $\sim_\alpha$ are a refinement of the equivalence classes of $\sim_\beta$. Another way of saying this is that if $\omega_1, \omega_2 \in \Sigma^*$ and $\omega_1 \sim_\alpha \omega_2$, then $\omega_1 \sim_\beta \omega_2$.

The map given by the recipe $[\omega]_\alpha \mapsto [\omega]_\beta$ is then well defined, and is a monoid epimorphism because of the trivial observation that for all $\omega_1, \omega_2 \in \Sigma^*$ we have $[\omega_1]_\beta[\omega_2]_\beta = [\omega_1\omega_2]_\beta$.

The only reason that we have introduced monoids is that the theory of presentations is most naturally set there.

Definition 6.7

Let Σ be a set of atoms, and let Σ^* denote the monoid of words on Σ. Suppose that R is a collection of equations of the form $l = r$ where $l, r \in \Sigma^*$. We write $\sim_R$ for the intersection of all translation invariant equivalence relations $\sim$ on Σ^* such that $l \sim r$ whenever $l = r$ is an equation in R. Now

$$mon\langle \Sigma \mid R \rangle$$

is a *monoid presentation* denoting the monoid of equivalence classes $\Sigma^* / \sim_R$. Informally we say that the monoid is defined on X subject to the relations R.

Note that a relation on Σ^* is just a subset of $\Sigma^* \times \Sigma^*$ and the intersection of equivalence relations is an equivalence relation. Moreover, it is straightforward to check that the intersection of translation invariant relations is translation invariant. We have the option of listing the elements of Σ instead of using the name of the set. Thus if $\Sigma = \{a, b\}$, we allow ourselves to write $mon\langle a, b \mid R \rangle$ instead of $\langle \Sigma \mid R \rangle$. Also if any of the equations has the form $w = 1$ where $w \in \Sigma^*$, then we allow ourselves to suppress the "=1" part of the equation. We can also take very modest liberties with equality, for example deeming $u = w = v$ to be short for the pair of equations $u = w$, $w = v$.

EXERCISES

6.1 Show that the monoid $M = mon\langle a \mid a^2 \rangle$ has two elements, and is a group.

6.2 Suppose that the equivalence relation $\sim_R$ is as given in Definition 6.7. Show that $u \sim_R v$ if and only if v can be obtained from u by finitely many replacements of a subword which is one side of an equation in R by the word which is the other side of the same equation.

6.3 Find the size of the monoid $mon\langle a \mid a^2 = a^3 \rangle$. Write down the multiplication table of this monoid.

We are building towards giving meaning to the group presentation $\langle X \mid R \rangle$. This looks very much like a monoid presentation, except that the equations in R can be between words which involve both elements of X and their (formal) inverses.

We make a shadow copy of X called X^- disjoint from X and distinguish a bijection "$-$" between these sets. We write x^- for the image of $x \in X$. We abusively also call the inverse bijection by the same name, so $(x^-)^- = x$. We now have the appropriate tools.

Definition 6.8

In the notation we have established, put $\Sigma = X \cup X^-$. Whenever $x \in X$ and x^{-1} occurs in an equation in R, replace x^{-1} by x^- and still call the collection of equations R. Let T be the collection of equations $xx^- = 1$ and $x^-x = 1$ as x ranges over X. Let

$$\langle X \mid R \rangle = mon\langle X \cup X^- \mid R \cup T \rangle.$$

Informally we say that we have presented a group X subject to the relations R.

The advantage of our construction is that we have defined a group presentation in terms of a monoid presentation, and the latter is easier to think about.

Definition 6.9

We use the established notation that $\Sigma = X \cup X^-$. A *translation invariant group equivalence relation* $\sim$ (or *tiger*[1] for short) is a translation invariant equivalence relation on Σ^* which has the property that $\sigma\sigma^- \sim \varepsilon$ for every $\sigma \in \Sigma$ (recall that ε denotes the empty word).

In the event that $\sim$ is a tiger, the set $\Sigma^*/\sim$ of equivalence classes is not just a monoid. It is a group because if $\omega = \sigma_1\sigma_2\ldots\sigma_n$ where each $\sigma_i \in \Sigma$, then $[\omega]^{-1} = [\sigma_n^-\sigma_{n-1}^-\ldots\sigma_1^-]$. Thus $\langle X \mid R \rangle$ is a group. Monoid homomorphisms between groups are group homomorphisms, so we deduce an important result.

Theorem 6.10 (Von Dyck, Prototype)

Suppose that $\sim_\alpha$ and $\sim_\beta$ are both tigers on $\Sigma = X \cup X^-$, and that $\sim_\alpha \subseteq \sim_\beta$, then the map which sends $[\omega]_\alpha \mapsto [\omega]_\beta$ is a group epimorphism.

Proof

This is simply Remark 6.6 in the context of groups.

□

Now suppose that H is a group and that $\Sigma = X \cup X^-$ where there is a distinguished bijection between the disjoint sets X and X^-. Suppose that $\theta : X \to H$ is any map. Extend the domain of θ first to Σ via $(x^-)\theta = ((x)\theta)^{-1}$

---

[1] This notation is home-made. It is not in general use.

for every $x \in X$, and then to Σ^* via $(\sigma_1 \sigma_2 \ldots \sigma_n)\theta = (\sigma_1)\theta \cdot (\sigma_2)\theta \cdot \cdots \cdot (\sigma_n)\theta$. Now we have an induced equivalence relation $\sim_H$ on Σ^* via $\omega_1 \sim_H \omega_2$ if and only if $(\omega_1)\theta = (\omega_2)\theta$. Notice that if $\omega_1 \sim_H \omega_2$ and $\lambda, \rho \in \Sigma^*$, then

$$(\lambda \omega_1 \rho)\theta = (\lambda)\theta \cdot (\omega_1)\theta \cdot (\rho)\theta = (\lambda)\theta \cdot (\omega_2)\theta \cdot (\rho)\theta = (\lambda \omega_2 \rho)\theta,$$

so $\lambda \omega_1 \rho \sim_H \lambda \omega_2 \rho$ and thus $\sim_H$ is translation invariant. Moreover $[\sigma_1 \sigma_2 \ldots \sigma_n]$ has $[\sigma_n^- \sigma_{n-1}^- \ldots \sigma_1^-]$ as a two-sided inverse. Thus $\sim_H$ is a tiger. There is a natural group monomorphism $[\omega]_H \mapsto (\omega)\theta$, and this is an isomorphism if θ (viewed as having domain Σ^*) is surjective.

We can now state Von Dyck's result in the way in which it is most widely recognized.

Theorem 6.11 (Von Dyck)

Suppose that X is a set and $\theta : X \to H$ is a map from X to a group H. Let $\langle X \mid R \rangle$ be a group presentation, and suppose that each relation $r \in R$ becomes a valid equation in H when the letters from X are replaced by their images in H. It follows that there is a unique group homomorphism $\overline{\theta} : \langle X \mid R \rangle \to H$ such that $([x]_R)\overline{\theta} = (x)\theta$ for all $x \in X$.

Proof

The map θ defines a tiger $\sim_H$ on Σ^* via extending θ to a homomorphism $\theta : \Sigma^* \to H$ as before. Now if $\alpha = \beta$ is in R, then $\alpha \sim_H \beta$ by hypothesis, so $\sim_R \subseteq \sim_H$. Now the prototype version of Von Dyck's theorem applies, and we have a unique homomorphism $\theta_1 : \Sigma^* / \sim_R \to \Sigma^* / \sim_H$ where $([\omega]_R)\theta_1 = [\omega]_H$, and we have a natural monomorphism $\Sigma^* / \sim_H \to H$ sending $[\omega]_H$ to $(\omega)\theta$. Composing we obtain a homomorphism $\overline{\theta} : \Sigma^* / \sim_R \to H$ which has the required property, and is clearly unique since the elements $[x]_R$ (where $x \in X$) generate $\Sigma^* / \sim_R$.

$\square$

Example 6.12

Here are three ways of writing the same group presentation (we are exhibiting how notation can vary):

$$\begin{aligned} G &= \langle a, b, c \mid a^2 = 1, b^3 = 1, c^5 = 1, ab = c^{-1} \rangle, \\ &= \langle a, b, c \mid a^2, b^3, c^5, ab = c^{-1} \rangle \text{ and} \\ &= \langle a, b, c \mid a^2 = b^3 = c^5 = 1, ab = c^{-1} \rangle. \end{aligned}$$

We are using shorthand in that b^3 is supposed to mean bbb. note that in the alternating group A_5 we have

$$(1,2)(3,4) \cdot (1,3,5) = (1,2,3,4,5).$$

By Von Dyke's theorem, the map $a \mapsto (1,2)(3,4)$, $b \mapsto (1,3,5)$, $c \mapsto (1,2,3,4,5)^{-1}$ extends to a unique homomorphism from G to A_5. In fact this is a surjection, as the reader may check.

6.3.1 Free Groups

At last we are in a position to make proper sense of the group F that we tried to introduce in Remark 6.3.

Definition 6.13

Let X be a set of atomic symbols. A group $F \supset X$ is said to be *free on* X if and only if whenever H is a group and $\theta : X \to H$ is a map, then there is a unique group homomorphism $\hat{\theta} : F \to H$ which extends the domain of θ.

First observe that $\langle X \mid - \rangle$ is a free group on X by Von Dyck's Theorem. Next the question of uniqueness arises. Suppose that F_1 and F_2 are both free groups on X, then the identity map $X \to X$ extends to two unique homomorphisms $F_1 \to F_2$ and $F_2 \to F_1$. The composites of these maps must be homomorphisms $F_1 \to F_1$ and $F_2 \to F_2$ which are both the identity on X. However, the only such maps are the identity maps again using the uniqueness clause. Our homomorphisms $F_1 \to F_2$ and $F_2 \to F_1$ are therefore mutually inverse isomorphisms.

Having established that F is unique in a very strong sense, we investigate its structure. Let $\Sigma = X \cup X^-$, then we say that a word $w = \sigma_1 \ldots \sigma_n$ ($\sigma_i \in \Sigma \forall i$) is freely reduced if it involves no juxtaposed pair $\sigma\sigma^-$ for $\sigma \in \Sigma$. Now, if $w = \lambda\sigma\sigma^-\rho$ for $\lambda, \rho \in \Sigma^*$, then in the free group we have $[w] = [\lambda\sigma\sigma^-\rho] = [\lambda\rho]$. Successive application of this process yields that every element of F is of the form $[u]$ where u is freely reduced. In fact it is also true that if u, u' are both freely reduced and $[u] = [u']$, then $u = u'$. Let Ω be the set of reduced words on Σ. Let $G = \mathrm{Sym}\ \Omega$ be the symmetric group on the set Ω (the set of all bijections from Ω to Ω). Define a map $\theta : X \to G$ via $w \cdot (x)\theta = wx$ if the last letter of w is not x^-. However, if $w = \lambda x^-$ is reduced as written, then let $w \cdot (x)\theta = \lambda$. It is easy to see that each $(x)\theta$ is in $\mathrm{Sym}\ \Omega$. By freeness, this map extends uniquely to a group homomorphism $\theta : F \to G$ where $(u)([w])\theta$ is calculated by writing the letters of w after u one at a time, and reducing (if appropriate) at each stage. If w is a reduced word on Σ^*, then $(1)(w)\theta = w$, so $\theta : F \to G$ is injective. It follows that each equivalence class $[v]$ contains exactly one irreducible word. Thus the irreducible words parameterize the elements of F.

All that was rather abstract. Let us prove a few down-to-earth facts about free groups. We will deliberately confuse words with their equivalence classes, but will use the subscript F to decorate the equals sign when indicating equality in the group F rather than simple equality of words. Thus if x is a generator of F then $x^{-1}x =_F 1$ is a correct equation since each side is the same group element. Moreover we will write x^{-1} instead of x^-.

Proposition 6.14

Suppose that F a free group on X. It follows that F is torsion-free (i.e. only the identity has finite order).

Proof

Suppose that $g \in F$ and g is not the identity. Thus $g = x_1 x_2 \cdots x_n$ where each x_i is a symbol in X or its inverse, the word $x_1 x_2 \cdots x_n$ is freely reduced and $n \geq 1$. We prove that g has infinite order by induction on n. If $n = 1$, then for each $m \in \mathbb{N}$ we have

$$g^m =_F \underbrace{x_1 \cdot x_1 \cdot \cdots \cdot x_1}_{m \text{ copies of } x_1}$$

and this expression is freely reduced so $g^m \neq_F 1$. If $n = 2$, then $x_1 x_n \neq_F 1$ so $x_n \neq x_1^{-1}$. Therefore for each $m \in \mathbb{N}$ we have

$$g^m =_F \underbrace{x_1 x_n \cdot x_1 x_n \cdot \cdots \cdot x_1 x_n}_{m \text{ copies of } x_1 x_n}$$

and this expression is freely reduced so $g^m \neq_F 1$.

When $n > 2$ there are two possibilities. Either $x_1 x_n \neq_F 1$ or $x_1 x_n =_F 1$. In the first case we argue as when $n = 2$. In the second case $g = x_1 h x_1^{-1}$ is freely reduced of length n so h is freely reduced of length $n - 2 \geq 1$. By induction on length, h has infinite order. Now g is a conjugate of h and so also has infinite order. $\square$

Free groups enjoy many properties. They are, for example, residually finite. We first define our tetrms.

Definition 6.15

Suppose that $\mathfrak{X}$ is a group property. Suppose that G is a group such that whenever $g \in G$ and $g \neq 1$, it follows that there is a group H_g with property $\mathfrak{X}$ and an epimorphism $\varphi_g : G \to H_g$ such that $(g)\varphi_g \neq 1$. Then we say that G is *residually* $\mathfrak{X}$.

Equivalently, if $\mathfrak{Y}$ is the collection of all normal subgroups N of G with the property that G/N has property $\mathfrak{X}$, then to say that G is a residually an $\mathfrak{X}$ group amounts to saying that

$$\bigcap_{N \in \mathfrak{Y}} N = 1.$$

If F is the free group on $X = \{x\}$, then $F = \langle x \rangle$ is infinite cyclic, and all of its subgroups are normal. Moreover

$$\bigcap_{m \in \mathbb{N}} \langle x^m \rangle = 1.$$

The groups $F/\langle x^m \rangle$ are finite cyclic groups. Therefore this particular free group is residually cyclic, residually abelian and residually finite. When X has more than one element, only one of these properties survives.

Proposition 6.16

Let F be a free group on an arbitrary set X. It follows that F is residually finite.

Proof

Suppose that

$$1 \neq_F g =_F x_1 x_2 \cdots x_n$$

where $x_1 x_2 \cdots x_n$ is a freely reduced word on the symbols from X and their inverses. Let $L_g = S_{n+1}$. To each $x \in X$ we will associate an element $\widehat{x} \in S_{n+1}$. If neither x nor x^{-1} arise as an x_i, then we put $\widehat{x} = \mathrm{id}$. Now for the interesting case where x or its inverse is involved at least once in g written as a reduced word. For each $i = 1, \ldots, n+1$ if $x_i = x$, then $\widehat{x} : i \mapsto i + 1$. Similarly for each $i = 1, \ldots, n+1$ if $x_i = x^{-1}$, then $\widehat{x} : i + 1 \mapsto i$. Since x cannot be juxtaposed with x^{-1} in a reduced word, no contradiction has been introduced. For each $x \in X$ involved in g we have a partial definition of an injective map from $\{1, 2, \ldots, n+1\}$ to itself. We extend each of these fragmentary definitions to produce n elements $\widehat{x}_i$ of S_{n+1}. Now $x_i \mapsto \widehat{x}_i$ for each i defines a map from X

to S_{n+1}. This extends uniquely to a homomorphism from F to S_{n+1}. Let H_g be the image of this map, and write $\varphi_g : G \to H_g$ for the induced epimorphism.

The map φ_g is so designed that

$$(1)(g)\varphi_g = (1)\widehat{x}_1 \widehat{x}_2 \cdots \widehat{x}_n$$

but $(i)\widehat{x}_i = i + 1$ for every i in the range $1 \le i \le n$, so $(1)(g)\varphi_g = n + 1 \ne 1$. Thus $(g)\varphi_g$ is not the identity element of the finite group H_g and we are done.

$\square$

Remark 6.17

This gives an alternative proof that different freely reduced words represent different elements of the free group F. To see this, suppose that w_1, w_2 are different freely reduced words. Let w be the reduced word obtained by freely reducing $w_1 * w_2^{-1}$. The boundary of the two words in the product is the only place where successive free reduction may occur. Note that w_2^{-1} is obtained by reversing the word w_2 and changing the sign of each exponent. The freely reduced word w is not empty since w_2^{-1} and w_1^{-1} are different freely reduced words (where w_1^{-1} is obtained from w_1 just as w_2^{-1} was obtained from w_2).

Now by Proposition 6.16 there is $N \trianglelefteq F$ with F/N finite and $Nw \ne N$. Therefore $w \ne_F 1$ so $w_1 =_F ww_2 \ne_F w_2$. The alternative proof is complete.

More delicate arguments will show that if p is a (fixed) prime number, then F is residually a finite p-group. All finite p-groups are nilpotent so and so every free group is residually nilpotent. There are many important and beautiful results about free groups with proofs which are beyond the scope of this text. For example, a subgroup of a free group is free. Another nice result is that if a torsion-free group G contains a free subgroup of finite index, then G is a free group. This generalizes Theorem 4.20 which deals with the case that X is a singleton set.

6.4 Presentations and Free Groups

Suppose that $H = \langle X \mid R \rangle$ and $F = \langle X \mid - \rangle$ and put $\Sigma^* = X \cup X^-$. We suppose that R consists of *relators* r_i standing for equations $r_i = 1$ rather than the more general relations $l_i = r_i$. Here i runs over an indexing set I. This has the advantage that each r_i (strictly speaking $[r_i]$) is actually an element of F so $R \subseteq F$. Let $N = \langle R \rangle^F$ denote the normal closure of R in F (recall Definition 2.16).

By Von Dyck's Theorem, the map $X \to \Sigma^*/ \sim_R$ defined by $x \mapsto [x]_R$ induces a group epimorphism $F \to H$. The kernel K of this map contains $[r_i]$ for every i, and thus $N \leq K$. Recall that this normal subgroup $N = \langle R \rangle^F$ of F is the intersection of all normal subgroups of F which contains every $[r_i]$, or equivalently it is the group generated by all conjugates of all the $[r_i]$.

Now, N defines an equivalence relation $\sim_N$ on Σ^* via $\omega_1 \sim_N \omega_2$ if and only if $[\omega_1]N = [\omega_2]N$. This is translation invariant since if $\lambda, \rho \in \Sigma^*$ and $\omega_1 \sim_N \omega_2$, then

$$[\lambda\omega_1\rho]N = [\lambda][\omega_1][\rho]N = [\lambda][\omega_1]N[\rho] = [\lambda][\omega_2]N[\rho] = [\lambda][\omega_2][\rho]N = [\lambda\omega_2\rho]N.$$

In fact this relation is also a tiger by design. Moreover, if $r_i \in R$, then $[r_i]N = [1]N$. Thus $\sim_R \subseteq \sim_N$. Now if $k \in K$, then $k \sim_R 1$ so $k \sim_N 1$ and therefore $kN = N$. We conclude that $k \in N$ and so $K \leq N$. We know $N \leq K$ so $N = K$.

The first isomorphism theorem applies and we see that we have an isomorphism between F/N and $H = \langle X \mid R \rangle$. This is a purely group theoretic construction. The isomorphism sends $N\omega$ to $[\omega]_R$ of course.

The reader may now recognize Von Dyck's theorem as another variation on the third isomorphism theorem.

6.4.1 Standard Sloppy Notation

We can now throw off Σ^* and take a purely group theoretic view. Instead of x^- we will write x^{-1}, and we will deliberately write w to mean the equivalence class $[w]$. When discussing presentations however, we may need to remind the reader that we are making this deliberate confusion. For example, if $G = \langle X \mid R \rangle$ and u, v are words on $(X \cup X^{-1})^*$, instead of writing $[u] = [v]$ we might write $u =_G v$ or $u = v$ in G. We might even write $u = v$, provided the meaning is absolutely clear from the context.

Theorem 6.18 (Von Dyck Revisited)

Von Dyck's theorem translates to say that any map $X \to Y$ induces a unique homomorphism $\langle X \mid R \rangle \to \langle Y \mid S \rangle$ providing that the image of each equation in R is an equational consequence of S (i.e. is valid in $\langle Y \mid S \rangle$).

6.4.2 Commutator Subgroups

We discussed the commutator subgroup (or derived group) at length in Chapter 5. We do not want the reader to have to revisit that material now, so we introduce the bare essentials once again.

Let H be a group. If $h, k \in H$, then we write $[h, k]$ for $h^{-1}k^{-1}hk$. Notice that $[h, k] = 1$ if and only if h and k commute. Let H' be the subgroup generated by all commutators (so if H is abelian, then $H' = 1$). Notice that for every $g \in H$ we have $g^{-1}[h, k]g = [h^g, k^g]$ so H' is normal in H. The group H/H' is obviously abelian, and moreover if M is normal in H and H/M is abelian then $H' \leq M \leq H$. Thus H' is the intersection of all normal subgroups of H yielding abelian quotients.

Suppose that $G = \langle X \mid R \rangle$. Let S be the set of relations consisting of equations $[x, y] = 1$ for all $x, y \in X$. We claim that (up to isomorphism) $\langle X \mid R \cup S \rangle$ is a presentation for G/G'. By Von Dyck's Theorem, the identity on X induces an epimorphism from $\langle X \mid R \cup S \rangle$ to G/G'. Also $\langle X \mid R \cup S \rangle$ is abelian (since generators commute) so the identity on X induces an epimorphism $G/G' \rightarrow \langle X \mid R \cup S \rangle$. These epimorphisms are mutually inverse, so each of them is an isomorphism, and thus $\langle X \mid R \cup S \rangle$ is a presentation for a group isomorphic to G/G'.

6.4.3 Tietze Transformations

Suppose that $G = \langle X \mid R \rangle$. There are *Tietze Transformations* which are manipulations of the presentation which do not change (up to isomorphism) the group presented. Moreover, it is possible to keep track of the isomorphism. These transformations are as follows.

(1) Suppose that y is a completely new atomic symbol, and w is any word in the letters of X and their inverses. The identity map on X induces an isomorphism $\langle X \mid R \rangle \rightarrow \langle X \cup \{y\} \mid R, w = y \rangle$

Proof of (1). By Von Dyck's Theorem, there is an induced homomorphism as suggested. There is also an induced homomorphism in reverse, being the identity on X and sending y to w. These maps are mutually inverse, and so are both isomorphisms.

(2) Suppose that $y \in X$ and let $X' = X - \{y\}$. Let $y = w$ be a relation and let the remaining relations comprise a set R'. Finally suppose that w is a word on $X' \cup X'^{-1}$ and that no relation other than $y = w$ involves the letter y or its inverse. The inclusion from X' to X induces an isomorphism from $\langle X' \mid R' \rangle$ to $\langle X \mid R \rangle$.

Proof of (2). Apply (1) to $\langle X' \mid R' \rangle$.

(3) The identity map on X induces an isomorphism from $\langle X \mid R \rangle$ to $\langle X \mid R \cup S \rangle$ where S is any set of equations on $X \cup X^{-1}$ which hold in the group $\langle X \mid R \rangle$.

Proof of (3). The extra equations do not change the tiger defined on $(X \cup X^{-1})^*$.

(4) One may reverse the process (3), just as one could reverse the process (2).

These Tietze transformations enable us to manipulate presentations without changing the group being defined. We illustrate the technique with an example. At each stage you should check which Tietze transformation is being applied. The cunning manipulation of presentations in this way is a very enjoyable black art.

Example 6.19

Let $G = \langle a, b \mid a^2 = 1, \; aba = b^{-1} \rangle$. This is actually isomorphic to the group of isometries of the integers given the usual metric. In fact G is a copy of the infinite dihedral group. The reader may check this as an exercise. Let $c = ab$ so

$$
\begin{aligned}
G &\simeq \langle a, b, c \mid a^2 = 1, \; aba = b^{-1}, \; c = ab \rangle \\
&\simeq \langle a, b, c \mid a^2 = 1, \; abab = 1, aba = b^{-1}, \; c = ab \rangle \\
&\simeq \langle a, b, c \mid a^2 = 1, \; abab = 1, \; c = ab \rangle \\
&\simeq \langle a, b, c \mid a^2 = 1, \; abab = 1, \; c = ab, c^2 = 1 \rangle \\
&\simeq \langle a, b, c \mid a^2 = 1, \; c = ab, c^2 = 1 \rangle \\
&\simeq \langle a, b, c \mid a^2 = 1, \; c = ab, c^2 = 1, b = a^{-1}c \rangle \\
&\simeq \langle a, b, c \mid a^2 = 1, \; c^2 = 1, b = a^{-1}c \rangle \\
&\simeq \langle a, c \mid a^2 = 1, \; c^2 = 1 \rangle.
\end{aligned}
$$

Chasing through the isomorphisms we see that a maps to a and b maps to $b = a^{-1}c$.

6.5 Decideability Problems

When doing geometry or topology, groups described by presentations crop up frequently. In order to make progress, it is sometimes important to discover the properties of the group $G = \langle X \mid R \rangle$.

   If both X and R are finite, then we say that $\langle X \mid R \rangle$ is a finite presentation of G. We say that G is *finitely presented* if it has at least one finite presentation. You might think that finitely presented groups are likely to be fairly easy to understand because they can be described by finite data. Alas this is not so. We run into decision problems normally associated with mathematical logic.

For example, suppose that $G = \langle X \mid R \rangle$ is a finite presentation. Now G is either the trivial group or it isn't. However, this issue cannot be resolved algorithmically. There is no mechanical procedure which can be let loose on an arbitrary finite presentation to resolve the matter in a finite number of steps. If G is the trivial group, then this can be verified by examining systematically all consequences of the relations, and discovering that $x =_G 1$ for every $x \in X$. We might let loose Robot A which has the task of verifying that G is trivial. If G is trivial, then this verification will finish successfully after a finite number of steps. However, if G is not trivial, then Robot A will slave away for eternity applying this algorithm without success. To the observer who does not know that G is non-trivial, the continuing work of Robot A means nothing. Either G is non-trivial, or G is trivial and if we have more patience Robot A will report the fact.

What we need, and what we cannot have, is Robot B which has an algorithm which will verify that G is non-trivial in finite time (if it is non-trivial!). Unfortunately no such algorithm exists.

If we have the extra information that G is finite, then Robot B has a much easier job. It can systematically list all groups of order n for each natural number n in turn via their Cayley tables, and test each possible map from G to each of these groups in turn to try to find an epimorphism. This can be done using Von Dyck's result because all we need to do is to see if the relations R hold in the finite group under scrutiny when we replace the letters of X by their images. If the map is a homomorphism, then we also need to check that the image is not trivial. If G is a non-trivial finite group (or even has a non-trivial finite homomorphic image), then this will be detected after a finite number of steps. This is a highly theoretical algorithm and is not at all speedy.

The problem is that there are finitely presented infinite groups which have no non-trivial finite images. For example, finitely presented infinite simple groups have this property.

Another question which cannot be resolved algorithmically is the *word problem*. Suppose that $G = \langle X \mid R \rangle$ is a finite presentation of the group G and that w_1, w_2 are words on the generators and their inverses regarded as elements of G. A solution to the *word problem* for G would be an algorithm which would take as input a presentation for G together with w_1, w_2 (words on the generators and their inverses regarded as elements of G) and decide in finitely many steps whether or not $w_1 =_G w_2$. It was shown by Post that there is a finitely presented monoid with an unsolvable word problem. There were considerable technical difficulties in generalizing Post's method to work for groups, but this was eventually achieved independently by Novikov in the Soviet Union and Boone in the United States. This happened in the 1950's at the height of the cold war, when communication between those countries was not easy.

Just as for the isomorphism problem, one of the verifications is algorithmic. If $w_1 =_G w_2$, then this can be algorithmically verified in finitely many steps. However, if $w_1 \neq_G w_2$, there is no algorithm which will infallibly detect this fact in a finite number of steps.

In general you cannot decide if a pair of elements of a finitely presented group are conjugate.

Do Not Despair

The fact that some questions one can ask about a finitely presented group are algorithmically undecideable is no reason to give up. After all, the groups which arise naturally when doing mathematics are rather special, and there is every reason to hope that we can understand a group which arises as an answer to a good question. The theory of *automatic groups* has recently been constructed. This class of groups includes many of the groups in which effective calculations can actually be performed, and has strong connections with hyperbolic geometry.

6.6 The Knuth–Bendix Procedure

The Knuth–Bendix procedure is an attempt to find a solution for the word problem in a presentation of an algebraic structure. It takes a particularly clean form for monoids, and the theory for monoids can then be deployed to groups. Thus we look at the monoid version of Knuth–Bendix.

Suppose we wish to be able to test for equality in $M = mon\langle \Sigma \mid R \rangle$ where Σ and R are both finite. We cannot expect to succeed in general because of Post's theorem. That does not stop us from trying.

Definition 6.20

A well-founded ordering $<$ of a set Ω is a partial order in which every strictly descending sequence is finite.

Well-founded translation invariant total orderings play such an important rôle in this story that they merit a special name.

Definition 6.21

Suppose that $M = mon\langle \Sigma \mid R \rangle$ is a monoid. A well-founded translation invariant total ordering of Σ^* is called a *reduction ordering*.

For the monoid M we might choose a total ordering of the finite alphabet Σ and then use the shortlex (length-then-lexicographic) ordering $<_{llex}$ (see Appendix B if necessary).

Example 6.22

A shortlex ordering on the words of a finite alphabet Σ is a reduction ordering on Σ^*. Each equivalence class of $\sim_R$, the equivalence relation induced by the relations R, will then contain a unique element of Σ^* minimal with respect to the ordering. Here, then, is a strategy for solving the word problem in M. Take two elements of the monoid $[\omega_1]$ and $[\omega_2]$ where $\omega_1, \omega_2 \in \Sigma^*$. Consider the minimal elements of these equivalence classes, say ω_1^{min} and ω_2^{min} respectively. Thus $[\omega_1] = [\omega_1^{min}]$ and $[\omega_2] = [\omega_2^{min}]$. The uniqueness of the minimal elements now forces $[\omega_1] = [\omega_2]$ if and only if $\omega_1^{min} = \omega_2^{min}$. These minimal words are elements of a free monoid, so they are equal if and only if they look equal.

Why have we failed to solve the word problem for monoids? The difficulty we have not yet addressed concerns the glib remark: *consider the minimal elements of the equivalence classes*. We know these elements exist, but in general there is no way to find them. The problem is that since we cannot solve the word problem, we cannot tell whether or not two words actually belong to the same equivalence class!

Suppose that $S \subseteq \Sigma^* \times \Sigma^*$. If $s \in S$ then we write $s = (i(s), t(s))$ where $i(s)$ is the *initial* of s and $t(s)$ is the *terminal* of s. We suppose that we have a reduction ordering $<$ on Σ^* and that for each $s \in S$ we have:

(i) $[i(s)] = [t(s)]$

(ii) $i(s) > t(s)$

(iii) If $\omega \in \Sigma^*$ is such that $\omega \neq \omega^{min}$, then there exists an element s of S such that $i(s)$ occurs as a subword of ω.

In this event we can obtain ω^{min} by using an algorithmic procedure providing that the set S is not too wild. Take ω, and scan it for a subword which is an initial of some $s \in S$. If no such subword is found, then $\omega = \omega^{min}$ by (3). If, on the other hand, $\omega = a \cdot i(s) \cdot b$ for some $a, b \in \Sigma^*$ and some $s \in S$, then set $\omega^{new} = a \cdot t(s) \cdot b$. By (i) we know

$$[\omega] = [a][i(s)][b] = [a][t(s)][b] = [\omega^{new}]$$

We replace ω by ω^{new} and go round the loop again. Now we use (ii); when passing through the loop, the translation invariance of the ordering ensures that $\omega > \omega^{new}$. The ordering, however, is well-founded, so after a finite number of passages through the loop we must find ω^{min}.

If S is finite the above process is algorithmic, but S need not be finite for this procedure to work. All we need is a way of testing whether or not a subword of ω is an initial of an element of S, and if it is, a way of calculating the appropriate terminal. Still, everything is very straightforward if S is finite.

A set S – whether it be finite or not – which satisfies (1),(2) and (3), is called a *complete set of rewrite rules*. We will explain the technical meaning of the word complete later, but for the moment we just understand it to mean "this does the job" in that successive application of the rules will put each word of an equivalence class into the same form. The expression *rewrite rule* should be clear enough. We rewrite ω as ω^{new} by replacing some subword $i(s)$ of ω by $t(s)$ for some $s \in S$.

6.7 Church–Rosser and Confluence

As usual we have an alphabet Σ and a relation on Σ^*. We will study the monoid $M = mon\langle \Sigma \mid R \rangle$. Now R consists of a collection of equations. These equations define a translation invariant equivalence relation $\sim_R$ on Σ^* which determines M.

Next suppose that we have a reduction ordering on Σ^*. Take each equation $\alpha = \beta$ in R and direct it into a rewrite rule $\alpha \to \beta$ or $\beta \to \alpha$, ensuring that the left-side is larger than the right-side in each rewrite rule (and discarding any equations of the form $\alpha = \alpha$). We write RW for the relation obtained by directing the equations of R in this way. Let us switch to infix notation, and write $\to$ for the translation closure of RW and $\to_*$ for the transitive closure of $\to$.

Definition 6.23

We say that RW is *locally confluent* if whenever $\omega \in \Sigma^*$ and both $\omega \to \mu$ and $\omega \to \nu$ then there exists some $\tau \in \Sigma^*$ such that both $\mu \to_* \tau$ and $\nu \to_* \tau$.

Each $s \in RW$ has the property that $i(s) > t(s)$, so life starts to look interesting. Given any $\omega \in \Sigma^*$, any sequence of application of rules

$$\omega \to \omega_1 \to \omega_2 \to \cdots$$

must terminate after a finite number of steps. We say that RW is a *terminating* set of rewrite rules.

Definition 6.24

We say that RW is *confluent* if whenever $\omega \to_* \mu$ and $\omega \to_* \nu$ then there exists $\tau \in \Sigma^*$ such that both $\mu \to_* \tau$ and $\nu \to_* \tau$.

Theorem 6.25 (Church–Rosser)

Using the notation above, if RW is locally confluent and terminating, then RW is confluent.

Proof

Let T (for trouble) stand for the set (which we hope is empty) of $\omega \in \Sigma^*$ such that there exist $\mu, \nu \in \Sigma^*$ with $\omega \to_* \mu$ and $\omega \to_* \nu$ but no $\tau \in \Sigma^*$ with $\mu \to_* \tau$ and $\nu \to_* \tau$. We assume (for contradiction) that T is non-empty. The ordering $<$ is well-founded, so there exists a minimal element of T. Let us call this ω_0. Suppose $\omega_0 \to \mu_0$ and $\omega_0 \to \nu_0$ are any one-step rewrites of ω_0 which extend to $\to_*$ rewrites of ω_0 to μ and ν respectively. By local confluence, there exists τ_0 such that $\mu_0 \to_* \tau_0$ and $\nu_0 \to \tau_0$. Now by the minimality of ω_0, there exist $\zeta_1, \zeta_2 \in \Sigma^*$ such that $\mu \to_* \zeta_1$, $\tau_0 \to_* \zeta_1$, and similarly $\nu \to_* \zeta_2$, $\tau_0 \to_* \zeta_2$. Now $\tau_0 < \omega$ so once again by the minimality of ω_0, there exists $\xi \in \Omega$ such that $\zeta_1 \to_* \xi$ and $\zeta_2 \to_* \xi$. Therefore both $\mu \to_* \xi$ and $\nu \to_* \xi$. This is impossible so T is empty, and RW is confluent.

$\square$

Now is the time for an important definition:

Definition 6.26

A terminating and confluent set of rewrite rules is called *complete*.

Notice that if $\omega \in \Sigma^*$ then there exists some ω_* such that $\omega \to_* \omega_*$ and ω_* cannot be rewritten. This follows from termination. Confluence forces ω_* to be the unique heir of ω with this property.

Definition 6.27

The element ω_* is called the *normal form* of ω and is written $nf(\omega)$.

Theorem 6.28

If $M = mon\langle \Sigma \mid R \rangle$ and RW is a complete set of rewrite rules, then in each $\sim_R$ class (these classes are the elements of the monoid) there is a unique $\omega \in \Sigma^*$ such that ω is not rewriteable; there is no μ with $\omega \to_* \mu$.

Proof

Define a relation $\sim$ on Σ^* by $\mu \sim \nu$ if and only if $nf(\mu) = nf(\nu)$. This is a translation invariant equivalence relation and is a subrelation of the one defining the monoid. Moreover, if $(\alpha, \eta) \in R$ then $nf(\alpha) = nf(\eta)$. Thus $\sim$ and $\sim_R$ coincide.

$\square$

This is all very good news. When we are given a monoid $\langle \Sigma \mid R \rangle$, we can select a reduction ordering on Σ^*, and direct the relations R according to the ordering to make them a set of rewrite rules RW. If our resulting rewrite rules are locally confluent, then they are actually confluent and we have an algorithm to solve the word problem. We also have the option, in the event that this fails to work, of rechoosing R. We can replace R by R' providing $mon\langle \Sigma \mid R \rangle = mon\langle \Sigma \mid R' \rangle$. This is exactly what the Knuth–Bendix procedure attempts to do. Adjoining new rewrite rules does not change the monoid, and the omission of a rule which is a consequence of other rules does not change the monoid either. If the Knuth–Bendix procedure halts, then the rules are locally confluent, and therefore (globally) confluent, and we have won.

Example 6.29

Let $F = \langle \Sigma \mid - \rangle$ be a free group. Well order the element of Σ in some way. Use the length then lexicographic ordering on Σ^*. The rules $aa^- \to 1$ and $a^- a \to 1$ (as a ranges over the letters of Σ) form a locally confluent (check!) set of rewrite rules for F. They therefore form a complete set of rewrite rules thanks to the Church-Rosser argument. Each element of F can be expressed by a unique unrewriteable word ω. If we indulge in the usual confusion, each word in Σ^* is equal to a unique "reduced word" in F (a word which contains no juxtaposition of a letter and its inverse). Thus we have another proof that distinct reduced words in a free group represent different group elements.

6.8 Normal Forms

We do not intend to dwell heavily on the subject of normal forms, but a few remarks are in order. Suppose $M = mon\langle \Sigma \mid R \rangle$ is a monoid and we have a complete set of rewrite rules RW which solve the word problem in M. The set RW may or may not have been obtained using the Knuth–Bendix procedure. The collection of normal forms will consist of all words which do not contain the initial of a rule as a subword. This is a very strong condition.

$$NF = \{\omega \in \Sigma^* \mid \not\exists \rho \in RW,\ \mu, \nu \in \Sigma*\ s.t.\ \omega = \mu i(\rho)\nu\}.$$

In terms of formal language theory, if RW is finite, then NF will be a regular language. In other words, there exists a finite-state automaton (FSA) which will recognize this language. We do not intend to develop this interesting topic, but merely to raise a little interest.

This ties in with the previous application. For the free group on, for example, the Roman alphabet, the shortlex ordering yields the set of normal forms which consists of words in which letters and their formal inverses are not juxtaposed i.e. the reduced words.

6.9 Knuth–Bendix for Strings

This procedure attempts to find a finite complete set of rewrite rules which will solve the word problem in a monoid. It is up to us to select a reduction ordering. We also have the option of finding another presentation for our monoid, and using that as our starting point. More accurately, we can find a presentation of an isomorphic copy of our monoid. We will work on a fixed presentation. Suppose $M = mon\langle \Sigma \mid R \rangle$ is a finite presentation of the monoid M (so both Σ and R are finite). We select a reduction ordering $<$ on Σ^*, and use the ordering $<$ to direct the equations R into rewrite rules $a_i \to b_i$ for all $i \in I$. In each case we must have $a_i > b_i$. A rule α has, as before, initial $i(\alpha)$ and terminal $t(\alpha)$.

It may be that the monoid we are studying is so amiable, and the ordering has been so judiciously chosen, that there exists a finite set of rewrite rules which will solve the word problem for us, and we go looking for such a set of rewrite rules.

The Knuth–Bendix procedure The input is a monoid presentation $M = mon\langle X \mid R \rangle$ where X and R are finite. We select a reduction ordering on X^*. We direct the equations into rewrite rules according by the larger side is rewritten to the smaller side. We then systematically test for local confluence using all possible overlaps between the left hand sides of rewrite rules. The point is that local confluence is only under threat when rewriting a word w if the two alternative applications of rewrite rules actually overlap in their application. Rather than give a formal definition of overlap, we hope the forthcoming example will clarify matters.

If ever there is a failure of local confluence (so w rewrites to distinct unrewriteable words w_1, w_2), then we know that $w_1 =_M w_2$. Direct this equation into a rewrite rule, append this rule to the list of rewrite rules we already have, and begin again. If ever we stumble across a locally confluent set of rewrite rules, then we stop, and by the Church–Rosser theorem, we have a complete set of rewrite rules for M.

Pairs of rules with initials which overlap in any non-empty way are called *Critical Pairs*.

Theorem 6.30

If the Knuth–Bendix procedure applied to a monoid presentation halts, the output is a complete set of rewrite rules which will solve the word problem in the monoid.

Proof

This is an application of Theorem 6.25

$\square$

Let us immediately demonstrate this with an example.

Example 6.31

Let $\Sigma = \{a, b, c\}$. We wish to study the monoid

$$M = mon\langle a, b, c \mid abc = a, \ bca = b\rangle.$$

We select the shortlex ordering on Σ^* where $a < b < c$. Our equations become rewrite rules:
Rules $= \{abc \to a, \ bca \to b\}$

We go hunting for critical pairs. The critical critical pairs between the rules involve
Critical $= \{(abc, bca), (bca, abc)\}$
Overlaps:$abca$ and $bcabc$).
Now $abca \to aa$ and $abca \to ab$ so we adjoin so adjoin a new rule $ab \to aa$.

Rules $= \{abc \to a, bca \to b, ab \to aa\}$
Critical$= \{(bca, abc), (ab, abc), (ab, bca)\}$
Overlaps: $(bcabc, abc, abca)$

Now we choose a critical pair to analyze: $bcabc \to bbc$ and $bcabc \to bca \to b$. We therefore adjoin the new rule $bbc \to b$.
Rules $= \{abc \to a, bca \to b, ab \to aa, bbc \to b\}$
Critical$= \{(ab, abc), (ab, bca), (bbc, bca), (ab, bbc)\}$
Overlaps: $(abc, abca, bbca, abbc)$ and round we go again: $abc \to aac$ and $abc \to a$ so we adjoin the new rule $aac \to a$.
Rules $= \{abc \to a, bca \to b, ab \to aa, bbc \to b, \ aac \to a\}$
Critical $= \{(ab, bca), (bbc, bca), (ab, bbc), (bca, aac)\}$
Overlaps:$(abca, bbca, abbc, bcaac)$.

Once more unto the breach. we have $abca \to aaca \to aa$ and $abca \to ab \to aa$. Perhaps the tide has turned. We can simply omit this critical pair.
Rules $= \{abc \to a, bca \to b, ab \to aa, bbc \to b, aac \to a\}$
Critical $= \{(bbc, bca), (ab, bbc), (bca, aac)\}$
Overlaps : $(bbca, abbc, bcaac)$.

Calculating with renewed vigour – we scent blood. We have $bbca \to ba$ and $bbca \to bb$. Oh dear, a new rule, but it is short and therefore very useful. $bb \to ba$.
Rules $= \{abc \to a, bca \to b, ab \to aa, bbc \to b, aac \to a, bb \to ba\}$
Critical$= \{(ab, bbc), (bca, aac), (bb, bca), (ab, bb), (bb, bbc)_1, (bb, bbc)_2\}$
Overlaps:$(abbc, bcaac, bbca, abb, bbc, bbbc)$.

Something peculiar is going on here. The initials bb and bbc overlap in two distinct ways. We have $abbc \to aabc \to aa$ and $abbc \to ab \to aa$ A winner! Keep **Rules** the same and feel happy.
Critical $= \{(bca, aac), (bb, bca), (ab, bb), (bb, bbc)_1, (bb, bbc)_2\}$
Overlaps: $(bcaac, bbca, abb, bbc, bbbc)$

Now we have $bcaac \to bac$ and $bcaac \to bca \to b$.
Rules $= \{abc \to a, bca \to b, ab \to aa, bbc \to b, aac \to a, bb \to ba, bac \to b\}$
Critical $= \{(bb, bca), (ab, bb), (bb, bbc)_1, (bb, bbc)_2, (ab, bac), (bb, bac)\}$
Overlaps: $(bbca, abb, bbc, bbbc, abac, bbac)$. We calculate again. We have $bbca \to baca \to ba$, $bbca \to bb \to ba$. Gold! Keep the rules the same.
Critical $= \{(ab, bb), (bb, bbc)_1, (bb, bbc)_2, (ab, bac), (bb, bac)\}$
Overlaps: $(abb, bbc, bbbc, abac, bbac)$.

The next calculation is $abb \to aab \to aaa$ and $abb \to aba \to aaa$. Thus we have
Critical $= \{(bb, bbc)_1, (bb, bbc)_2, (ab, bac), (bb, bac)\}$
Overlaps: $(bbc, bbbc, abac, bbac)$.

We have $bbc \to bac \to b$ and $bbc \to b$.
Critical $= \{(bb, bbc)_2, (ab, bac), (bb, bac)\}$

Overlaps: $(bbbc, abac, bbac)$.

Next we see that $bbbc \to babc \to ba$ and $bbbc \to bb \to ba$.

Critical $= \{(ab, bac), (bb, bac)\}$

Overlaps:$(abac, bbac)$

We see that $abac \to aaac \to aa$ and $abac \to ab \to aa$.

Critical$= \{(bb, bac)\}$

Overlap:$(bbac)$.

We have $bbac \to baac \to ba$ and $bbac \to bb \to ba$ so the procedure has terminated.

Final Rules

$abc \to a$, $bca \to b$, $ab \to aa$, $bbc \to b$, $aac \to a$, $bb \to ba$ and $bac \to b$.

We may now omit redundant rules to obtain reduced set of final rules. If a rule need never be used (because another rule has a left side which is a subword of its left hand side), then the rule serves no purpose.

Reduced Final Rules

(1) $bca \to b$

(2) $ab \to aa$

(3) $aac \to a$

(4) $bb \to ba$

(5) $bac \to b$

These rules form a complete rewriting system for the monoid M. We may use them to solve the word problem in the monoid. For example, someone might ask whether or not $[bacbac]$ was equal to $[ababaccccb]$. Without access to a complete set of rewrite rules, this query would be extremely vexing. We however, have the means to solve this problem in a trice. Now

$$bacbac \to_{(1)} bbac \to_{(1)} bb \to_{(4)} ba$$

whereas

$$ababaccccb \to_{(2)} aaabaccccb \to_{(2)} aaaaaccccb \to_{(3)} aaaaccccb \to_{(3)}$$

$$aaaccb \to_{(3)} aacb \to_{(3)} ab \to_{(2)} aa$$

and so $bacbac$ and $aaabaccccb$ are not equal in our monoid M.

EXERCISES

6.4 In fact the ordering we chose on Σ^* was quite arbitrary. Try finding a complete set of rewrite rules to solve the word problem in M using length-then-lexicographic ordering, but with the alphabet ordered via $b < a < c$. This does not take long.

7
Appendix A: Fields

7.1 The Axioms

Informally, a field is a set equipped with addition, subtraction, multiplication and division in which all the laws of algebra taught in secondary schools are valid. That sentence was vague beyond belief, but gives the spirit of what is going on.

The Field Axioms. More formally then, a *field* is a set F containing distinguished elements 0 and 1 and which comes equipped with closed binary operations $+$ and $\times$ such that a list of axioms are satisfied.

(a) F is an abelian group under $+$ (addition), with identity element 0.

(b) The operation $\times$ (multiplication) is associative, commutative, distributes over addition, and 1 acts as a multiplicative identity.

(c) $F - \{0\}$ is a group under multiplication.

We elaborate a little. Since multiplication is commutative we do not have to worry about one-sided distributivity or 1 being a one-sided multiplicative identity. Axiom (b) says that

$$(x \times y) \times z = x \times (y \times z) \; \forall x, y, z \in F,$$

and that

$$x \times (y + z) = (x \times y) + (x \times z) \; \forall x, y, z \in F.$$

It goes on to specify that

$$x \times y = y \times x \ \forall x, y \in F$$

and that

$$1 \times x = x \times 1 = x \ \forall x \in F.$$

Examples of fields include the rational numbers $\mathbb{Q}$, the real numbers $\mathbb{R}$ and the complex numbers $\mathbb{C}$. However, these are merely the best known examples of fields. There are also fields of finite size.

Before we dispense with the formalities, we should really check that the field axioms do give us everything we want. Subtraction and division are not explicitly mentioned in the field axioms, but they are easy operations to build. If $x, y \in F$ where F is a field, let $x - y$ be defined to be $x + (-y)$ where $(-y)$ is the additive inverse of y. Moreover, if $y \neq 0$, then y is an element of the multiplicative group $F - \{0\}$. Thus it has a unique multiplicative inverse y^{-1}, and we can define $x \div y$ to be xy^{-1}.

Various plausible statements about fields are true because of axiom chasing. For example

Lemma 7.1

Let F be a field. Suppose that $x \in F$, then $x \times 0 = 0$.

Proof

We know that $0 + 0 = 0$ since 0 is the additive identity of F. Therefore $x \times (0 + 0) = x \times 0$ so by distributivity we have

$$(x \times 0) + (x \times 0) = x \times 0.$$

Now add the additive inverse of $x \times 0$ to each side of this equation, so that

$$((x \times 0) + (x \times 0)) + (-(x \times 0)) = x \times 0 + (-(x \times 0)).$$

Now use associativity of addition on the left side of this equation, and the definition of additive inverse on the right side to obtain

$$(x \times 0) + ((x \times 0) + (-(x \times 0))) = 0.$$

Next use the definition of additive inverse to obtain $(x \times 0) + 0 = 0$ but 0 is the additive identity so $x \times 0 = 0$.

$\square$

Here is another likely looking result.

Lemma 7.2

Let F be a field. Suppose that $x, y \in F$ and that $x \times y = 0$. It follows that $x = 0$ or $y = 0$.

Proof

This follows because $F - \{0\}$ is a multiplicative group, and is therefore closed under multiplication. If $x \neq 0 \neq y$, then $xy \in F - \{0\}$ and so $xy \neq 0$.

$\square$

We have used brackets (parentheses) to indicate priority of operations. However, there are well-established conventions allowing you to omit brackets and yet be well-understood. For example $a \times b + c$ means $(a \times b) + c$ rather than $a \times (b + c)$. We will adhere to these conventions from now on. Also $\times$ is a little cumbersome. When doing algebra it is often possible to use mere juxtaposition to indicate multiplication, so xy is short for $x \times y$. Sometimes the context does not allow this; for example, it would be dangerous to multiply the natural numbers 6 and 7 by writing $67 = 42$. In such circumstances, a central or lower dot may suffice, so $6 \cdot 7 = 42$ and $6.7 = 42$ are both possible notations. We will rarely if ever have recourse to decimal notation so there should be no ambiguity.

7.2 Finite Fields

Definition 7.3

Let F be field such that $|F| < \infty$. We say that F is a *finite field* or *Galois field* and that the *order* of F is $|F|$.

Let n be a natural number (and for us, $0 \notin \mathbb{N}$). The set $\langle n \rangle$ of integer multiples of n is an additive subgroup of $\mathbb{Z}$. We can form the additive factor group $\mathbb{Z}/\langle n \rangle$ which we write as $\mathbb{Z}_n$. We define a multiplication on $\mathbb{Z}_n$ via $(x + \langle n \rangle) \cdot (y + \langle n \rangle) = xy + \langle n \rangle$ whenever $x, y \in \mathbb{Z}$. We must verify that this definition is legitimate. The problem is that it is possible that $x' + \langle n \rangle = x + \langle n \rangle$ even though the integers x, x' are distinct. Now suppose that $x', y' \in \mathbb{Z}$, and that

$x' + \langle n \rangle = x + \langle n \rangle$ and $y' + \langle n \rangle = y + \langle n \rangle$. Then n divides both $x' - x$ and $y' - y$. Thus

$$x'y' - xy = (x' - x)y' - x(y - y') \in \langle n \rangle.$$

It follows that $xy + \langle n \rangle = x'y' + \langle n \rangle$ and so this operation is well-defined. Note that $\mathbb{Z}_n$ is an additive group, and the multiplication we have defined is associative, commutative and $1 + \langle n \rangle$ is a (and therefore the) multiplicative identity. Multiplication distributes over addition in $\mathbb{Z}_n$. For any $x \in \mathbb{Z}$ we write $\mathbf{x}$ as economical notation for $x + \langle n \rangle \in \mathbb{Z}_n$.

Proposition 7.4

The structure $\mathbb{Z}_n$ defined above is a field if and only if the natural number n is prime.

Proof

Certainly most of the field axioms are valid in $\mathbb{Z}_n$ for any natural number n. The coset $\mathbf{0} = 0 + \langle n \rangle$ will be the additive identity, and $\mathbf{1} = 1 + \langle n \rangle$ will be the multiplicative identity. If n is not prime, then $n = 1$ or $n = rs$ is composite, with $r, s \in \mathbb{N}$ and $1 < r, s < n$. In the first instance $\mathbb{Z}_1$ has just one element, so the set of non-zero elements of $\mathbb{Z}_1$ is empty and therefore cannot be a group. Thus Axiom (b) does not hold.

If n is composite then $\mathbf{r} \neq \mathbf{0} \neq \mathbf{s}$, but $\mathbf{rs} = \mathbf{n} = \mathbf{0}$. By Lemma 7.2 the structure $\mathbb{Z}_n$ is not a field.

Now we address the positive side. If $n = p$ is a prime number, then $p \geq 2$ so

$$1 + \langle p \rangle = \mathbf{1} \neq \mathbf{0} = 0 + \langle p \rangle$$

so $\mathbb{Z}_p - \{\mathbf{0}\}$ is not empty.

It remains to show that $\mathbb{Z}_p - \{\mathbf{0}\}$ is a multiplicative group. If $\mathbf{x}, \mathbf{y} \in \mathbb{Z}_p - \{\mathbf{0}\}$ then $\mathbf{x} = x + \langle p \rangle$ and $\mathbf{y} = y + \langle p \rangle$ for integers x and y. Now p divides neither x nor y, so p does not divide xy and so $\mathbf{xy} \neq \mathbf{0}$. Thus $\mathbb{Z}_p - \{\mathbf{0}\}$ is closed under multiplication.

Next we address for multiplicative inverses. Choose any $\mathbf{x} = x + \langle p \rangle$ a non-zero element of $\mathbb{Z}_p$. The greatest common divisor of x and p is 1 since p is not a divisor of x. Therefore (by Euclid's algorithm) there are integers r, s such that $rx + sp = 1$. Thus $rx - 1 \in \langle p \rangle$, so $\mathbf{rx} = \mathbf{1}$. We conclude that $\mathbf{r}$ is a multiplicative inverse for $\mathbf{x}$.

□

We have shown that for each prime number p, there is a field $\mathbb{Z}_p$ which has p elements. Now suppose that F is another field with p elements. Now additively both $\mathbb{Z}_p$ and F are cyclic groups of prime order p, and each of the multiplicative identities will do as an additive cyclic generator. There is exactly one additive group isomorphism from $\theta : \mathbb{Z}_p \to F$ which maps $\mathbf{1}$ to 1_F. We claim that θ respects the multiplicative structure. If $\mathbf{x}, \mathbf{y} \in \mathbb{Z}_p$ we may assume that $\mathbf{x} = [x]$ and $\mathbf{y} = [y]$ with $x, y \in \mathbb{N}$. Now $\mathbf{x} = \sum_{i=1}^{x} \mathbf{1}$ and $\mathbf{y} = \sum_{j=1}^{y} \mathbf{1}$ so

$$\mathbf{x}\mathbf{y} = \left(\sum_{i=1}^{x} \mathbf{1}\right)\left(\sum_{j=1}^{y} \mathbf{1}\right) = \sum_{k=1}^{xy} \mathbf{1}.$$

Now $\theta : \mathbf{x} = \sum_{i=1}^{x} \mathbf{1} \mapsto \sum_{i=1}^{x} 1_F$, $\theta : \mathbf{y} = \sum_{i=1}^{y} \mathbf{1} \mapsto \sum_{i=1}^{y} 1_F$, and

$$\theta : \mathbf{x}\mathbf{y} = \sum_{i=1}^{xy} \mathbf{1} \mapsto \sum_{i=1}^{xy} 1_F.$$

Therefore $(\mathbf{x}\mathbf{y})\theta = (\mathbf{x})\theta(\mathbf{y})\theta$. The map θ is a bijection which respects both the additive and multiplicative structures, and therefore deserves to be called an *isomorphism of fields* (the reader is invited to check that the inverse of a field isomorphism is an isomorphism).

Thus, up to isomorphism, there is a unique finite field of size p. It is sometimes called GF(p) (the *Galois field* with p elements). The term Galois field is a synonym for finite field, and the terminology pays tribute to the celebrated brilliant young mathematician who died tragically young.

Now suppose that F is any finite field. Consider the additive subgroup $S = \langle 1_F \rangle$ of F generated by 1_F. Notice that S is closed under multiplication. The size of S is the least natural number n such that 1 added to itself n times is 0. It follows that n is a prime number p, else we can argue as above and produce two non-zero elements of F with product 0. The multiplicative structure of S is determined by

$$\left(\sum_{i=1}^{x} 1_F\right)\left(\sum_{i=1}^{y} 1_F\right) = \sum_{i=1}^{xy} 1_F$$

and this is exactly the same recipe as in $\mathbb{Z}_p$. Therefore S is a field and S is a copy of GF(p). When a field F contains a subfield which is a copy of GF(p) we say that F is a field of *characteristic p*. If F contains no copy of any GF(p) then it contains a copy of the integers, and therefore of the rational numbers, and we say that F has characteristic 0.

Suppose that F is a finite field, then F is a vector space of dimension n over S where $n \in \mathbb{N}$. Let $u_1, \ldots, u_n$ be an ordered basis of F as an S-vector space. Every element of F is uniquely expressible as $\sum_{i=1}^{n} s_i u_i$ for suitable $s_i \in S$. Therefore $|F| = p^n$. We have proved that any Galois field F must have a prime-power number of elements.

We introduce some standard terminology. Let F be a field an X be an "variable". The set of all polynomials in X with coefficients in F is written $F[X]$. Note that the field elements, including 0, are all elements of $F[X]$ since they are the constant polynomials. We will assume a casual familiarity with the algebra of polynomials, including knowledge of the function degree : $F[X] \rightarrow \mathbb{N} \cup \{0\} \cup \{-\infty\}$. Note that the degree of a non-zero constant is 0 but the degree of 0 is $-\infty$. This convention allows the statement of simple truths concerning degree without having to mutter "except for the zero polynomial" under your breath. If $f, g \in F[X]$, then $\deg (f + g) \leq \max\{\deg f, \deg g\}$ and $\deg fg = \deg f + \deg g$. If $f \in F[X]$, then we sometimes write f as $f(X)$ to allow evaluation or other substitution of the variable. We assume knowledge of roots and evaluation.

Lemma 7.5

Let F be a field. Suppose that $f \in F[X]$ is a non-zero polynomial and that $\alpha \in F$. It follows that there is a polynomial q such that

$$f = q \cdot (X - \alpha) + f(\alpha).$$

Proof

Let $h(X) = f(X + \alpha) \in F[X]$. Now

$$f = f(X) = h(X - \alpha) = q(X) \cdot (X - \alpha) + c.$$

Evaluating each side at α it follows that $c = f(\alpha)$ and we are done.

$\square$

Corollary 7.6

Suppose that $f \in F[X]$ is a non-zero polynomial and that $\alpha \in F$ is a root of F. It follows that there is a polynomial $q \in F[X]$ such that $f = q \cdot (X - \alpha)$.

Proposition 7.7

Let F be a field, and let $f \in F[X]$ be a non-zero polynomial. It follows that the number of distinct roots of f in F is at most $\deg f$.

Proof

We induct on the degree of f, a non-negative integer. If the degree of f is 0, then f has no roots in F. If the degree of f is greater than 0, either f has no roots in F (and we are done) or it has a root α. Now $f = q \cdot (X - \alpha)$ for a non-zero polynomial q by Lemma 7.5, and q has at most $(\deg f) - 1$ roots in F by inductive hypothesis. Thus f has at most $\deg f$ roots in F.

$\square$

In some expositions the notion of formal differentiation is introduced in connection with the occurrence of multiple roots of polynomials. In the proof of the next result we use a trick which avoids the need for this artifice.

Proposition 7.8

Let F be a finite field of order p^n, then $X^{p^n} - X = \prod_{\alpha \in F}(X - \alpha)$.

Proof

Certainly 0 is a root of $X^{p^n} - X$. Moreover, if $\alpha \in F - \{0\}$, then $\alpha^{p^n - 1} = 1$ since $F - \{0\}$ is a multiplicative group and Lagrange's theorem applies. Thus $\alpha^{p^n} = \alpha$ so each $\alpha \in F$ is a root of $X^{p^n} - X$.

Now $h = X^{p^n} - X - \prod_{\alpha \in F}(X - \alpha)$ has degree less than p^n and has p^n roots. Now Proposition 7.7 applies and we deduce that h is the zero polynomial and so $X^{p^n} - X = \prod_{\alpha \in F}(X - \alpha)$.

$\square$

Proposition 7.9

Let F be a Galois field. It follows that the multiplicative group $F^* = F - \{0\}$ is cyclic.

Proof

If $|F| = 2$, then F^* is trivial and therefore cyclic. From now on we assume that $|F| > 2$ so F^* is not the trivial group. Suppose (for contradiction) that F^* is not cyclic. Therefore

$$F^* \simeq C_{d_1} \times C_{d_2} \times \cdots \times C_{d_t}$$

with $t > 1$, and $1 < d_1, \ldots, d_t$ natural numbers with the property that d_i divides d_{i+1} for each $i < t$. This follows from Theorem 2.44. Now each factor in this Cartesian product has a unique cyclic subgroup of order d_1, so F contains d_1^t elements α such that $\alpha^{d_1} = 1$. Now the polynomial $X^{d_1} - 1$ has at least d_1^t roots in F. Now Proposition 7.7 applies and we deduce that $t = 1$.

$\square$

Proposition 7.10

Let F be a field of order p^n where p is a prime number. The map $\varphi : F \to F$ defined by $f \mapsto f^p$ is an isomorphism of fields.

Proof

First note that the subfield generated by adding 1 to itself must have order p. Therefore if $f \in F$ is added to itself $p - 1$ times then

$$\underbrace{f + f + \cdots + f}_{p \text{ copies of } f} = \underbrace{(1 + 1 + \cdots 1)}_{p \text{ copies of } 1} f = 0 \cdot f = 0.$$

Certainly φ respects multiplication. Addition is more interesting. The important point is that each binomial coefficients $\binom{p}{i}$ (an entry of Pascal's triangle) is an integer divisible by p whenever $1 \leq i \leq p - 1$. This is because

$$\binom{p}{i} = \frac{p!}{(p - i)!i!} \in \mathbb{Z}$$

and the act of cancelling the denominator to 1 will leave the factor of p in the numerator untouched.

Therefore if $a, b \in F$, then

$$(a + b)^p = a^p + \sum_{i=1}^{p-1} \binom{p}{i} a^i b^{p-i} + b^p = a^p + b^p.$$

The point is that the notation $\binom{p}{i} a^i b^{p-i}$ means the result of summing $\binom{p}{i}$ copies of $a^i b^{p-i}$. Now the result of summing just p copies of $a^i b^{p-i}$ is 0 so the result of summing $\binom{p}{i}$ copies will also be 0. Therefore

$$(a + b)^p = a^p + b^p$$

for each $a, b \in F$. The map φ is an additive group homomorphism with trivial kernel. It is therefore injective. Since it is a map from a finite set to itself it is bijective.

$\square$

Corollary 7.11 (of the Proof)

Suppose that F is a (possibly infinite) field containing $\mathbb{Z}_p$ as a subfield. The map $f \mapsto f^p$ is a monomorphism.

Proposition 7.12

Let F be a field and consider the polynomial $f \in F[X]$. It follows that there is a field K with F a subfield of K such that f is a product of linear polynomials in $K[X]$.

The reader should expand upon the following outline.

Proof

We factor f into irreducible (unfactorizable) elements of $F[X]$, and induct on the sum of the degrees of the non-linear polynomials in our factorization. If our expression of f a product irreducible polynomials involves only linear polynomials, then put $K = F$. If there is a non-linear polynomial g in the product we make a construction.

Change the variable from X to t, and temporarily regard g as a polynomial in t. Define an equivalence relation on $F[t]$ via $u(t) \sim v(t)$ if and only if there is $q(t) \in F[t]$ such that $u(t) - v(t) = q(t) \cdot g(t)$. It is routine to check that $\sim$ is an equivalence relation. Moreover, it is easy to check that addition and multiplication in $F[t]$ induce multiplicative and additive structures on the set $L = F[t]/\sim$ of equivalence classes. Moreover, L is a field (checking this is quite a lot of work). The map from F to L defined by $a \mapsto [a]$ (the equivalence class of the constant polynomial a) is an embedding (a monomorphism) of the field F into the field L. Identifying F with its image, we may think of F as a subfield of L. (It also follows that $\dim_F(L) = \deg g$).

Now let $\alpha = [t]$ be the equivalence class of t and revert to thinking of g as a polynomial in the variable X. We evaluate $g(\alpha)$ by

$$g(\alpha) = g([t]) = [g(t)] = [0] = 0_L$$

since $g(t) \sim 0$. Therefore $\alpha \in L$ is a root of g and so by Corollary 7.6 $g = (X - \alpha)g'$ for $g' \in L[X]$. We have reduced the sum of the degrees of the non-linear irreducible factors of f by working in $L[X]$. We replace K by L and finish by induction.

$\square$

Proposition 7.13

For each prime number p and for each natural number n there is a field M of order p^n.

Proof

Let F be the field with p elements and put $h = X^{p^n} - X \in F[X]$. Using Proposition 7.12 we conclude that there is an overfield K of F such that h factorizes into linear polynomials in $K[X]$. Let M denote the subset of K consisting of the roots of h. It is a routine matter to verify that M is a field and that F is a subfield of M. It follows that $|M| = p^m$ where $m \leq n$.

Suppose that α is a root of h, then

$$h(X) = h(X - \alpha + \alpha) = h(X - \alpha) + h(\alpha) = h(X - \alpha).$$

The second of the three equalities follows from the fact that raising elements to the power p is an automorphism (self-isomorphism) of M. Now

$$h(X - \alpha) = (X - \alpha)(X - \alpha)^{p^n - 1} - (X - \alpha)$$

and so α is not a root of $h/(X - \alpha)$. The roots of h are therefore distinct and so $|M| = p^n$.

$\square$

A more careful analysis will show that if N is any field containing p^n elements then M and N are isomorphic fields. Thus up to isomorphism, there is a unique field of each prime power order.

8

Appendix B: Relations and Orderings

Let Ω be a non-empty set. A relation on Ω is simply a subset of the Cartesian square of Ω. In symbols we could write $R \subseteq \Omega \times \Omega$. Suppose that $(a, b) \in R$. A very suggestive alternative way to express this is to use infix notation and write aRb to mean $(a, b) \in R$. For some reason, probably habit, people seem to prefer some non-alphabetic symbol when using infix notation. Let us pander to this, and write $a \sim b$ as yet another synonym for $(a, b) \in R$. If Ω_1 is a non-empty subset of Ω then $R \cap (\Omega_1 \times \Omega_1)$ is the *induced relation* on Ω_1, though it usually is still called R (the benign supposition being that the set is evident from the context).

If R_1, R_2 are two relations on Ω with $R_1 \subseteq R_2$ we say that R_1 is a subrelation of R_1. We should also be prepared to say that R_2 is an *overrelation* of R_2 – though the word overrelation is rather ugly.

An arbitrary relation is considerably less interesting than watching paint dry. The relation must be in some way special in order to be worthy of attention. Here are just a few of the properties which a relation might have.

(a) For each $\omega \in \Omega$ there exists a unique ν in Ω such that $\omega \sim \nu$. In this case the relation is really a **map** from Ω to Ω.

(b) For each $\omega \in \Omega$ we have $\omega \sim \omega$. We then say $\sim$ has the **reflexive** property.

(c) Whenever $\omega, \nu \in \Omega$ are such that $\omega \sim \nu$ then also $\nu \sim \omega$. In this case we say $\sim$ has the **symmetric** property.

(d) Whenever $\omega, \nu, \mu \in \Omega$ are such that both $\omega \sim \nu$ and $\nu \sim \mu$ then it follows that $\omega \sim \mu$. This time we say that $\sim$ has the **transitive** property.

A relation satisfying (b),(c) and (d) simultaneously is called an **equivalence relation.** We assume that the reader is familiar with the notion of an **equivalence class.** In the event that $\sim$ is an equivalence relation we use the symbol $\Omega/\sim$ to denote the set of equivalence classes. In this context a subset T of Ω with the property that $T \cap C$ is a singleton set for each $C \in \Omega/\sim$ is called a transversal for the equivalence classes. In happy circumstances, Ω will possess some sort of algebraic structure, and this will be inherited by the equivalence classes. It may or may not be possible to choose a transversal which captures this algebraic structure.

Another important class of relations on Ω are the partial orders. We shall postpone a discussion of these until later.

Closure

Let $S = \Omega \times \Omega$. This is certainly an equivalence relation, and for infix purposes we will denote it by $\sim_{uni}$. This is the universal equivalence relation in the sense that $\omega_1 \sim_{uni} \omega_2$ for all $\omega_1, \omega_2 \in \Omega$. Note that S satisfies (b), (c) and (d).

Now suppose that R is an arbitrary relation on Ω. Thus $R \subseteq S$ so R is contained in a relation which is simultaneously reflexive, symmetric and transitive. Notice that the intersection of relations satisfying (x) (where x is b, c or d) is also a relation satisfying (x). Thus there is a smallest reflexive (or symmetric, or transitive) relation containing R. We call these three relations the *reflexive closure*, the *symmetric closure* and the *transitive closure* of R respectively. Similarly we can consider the intersection of all relations containing R which simultaneously satisfy (b),(c) and (d). In other words, we can look at the intersection of all the equivalence relations containing R. This itself will be an equivalence relation and is termed the *equivalence closure* of R.

So much for existence. We now investigate a more direct method of getting at these closures. We first examine the transitive closure of R which we call TR or $\sim_{TR}$. Thus $R \subseteq TR$. If we have a finite collection of R-related pairs $\omega_i \sim \omega_{i+1}$ where $i = 0, \ldots n-1$ then it follows that $\omega_0 \sim_{TR} \omega_n$. We can twist this around to get at TR directly. Define a new relation $\sim_{new}$ as follows: write $\nu \sim_{new} \mu$ if and only if there is a finite collection of R-related pairs $\omega_i \sim \omega_{i+1}$ with $i = 0, \ldots, n-1$ and $\omega_0 = \nu$, $\omega_n = \mu$. This relation on Ω is visibly transitive, and contains R as a subrelation. Moreover the relation $\sim_{new}$ is contained in TR (or $\sim_{TR}$ if you prefer). Now, $\sim_{TR}$ is the smallest transitive relation containing R, and so $\sim_{new} = \sim_{TR}$. We now have a concrete way of viewing $\sim_T R$.

The devices that enable us to construct the reflexive and symmetric closures of R are even easier. To get the reflexive closure of R you just adjoin all pairs

(ω, ω) to R. To get the symmetric closure of R, first define a new relation (the *opposite* of R) by setting $R^{op} = \{(\nu, \omega) \mid \omega \sim \nu\}$. The symmetric closure of R is just $R \cup R^{op}$. The reader is invited to justify the assertions of this paragraph, and to do the following exercises.

EXERCISES

Let R be a relation on a set X.

8.1 Show that the equivalence closure of R is the transitive closure of the symmetric closure of the reflexive closure of R.

8.2 Why cannot we define the *map closure* of an arbitrary relation R? That is, why cannot we do for (a) what we have done for (b), (c) and (d)?

8.3 *An Old Chestnut:* When defining an equivalence relation, why do we need to assume that (a) holds? Surely we can deduce (a) from (b) and (c), or perhaps we cannot? Discuss.

8.4 Suppose $|\Omega| = n < \infty$. How many relations are there on Ω? Let Σ denote the set of relations on Ω. How many relations are there on Σ?

8.1 Orderings

We are all familiar with the usual ordering on the integers $\mathbb{Z}$. There is a binary relation $<$ on $\mathbb{Z}$ such that, for example, $-1 < 8$ and $42 < 666$. In order to make absolutely sure of what we mean by an ordering, a definition is called for:

Definition 8.1

A relation $<$ on a set Ω is said to be an *ordering* if and only if the following two conditions are satisfied:

(i) If $a, b, c \in \Omega$ are such that $a < b$ and $b < c$ then $a < c$.

(ii) If $a, b \in \Omega$ and $a < b$ then it is not the case that $b < a$.

We may describe the pair $(\Omega, <)$ as an *ordered set*. Sometimes it is convenient to reverse the ordering symbol. We write $b > a$ as alternative notation for $a < b$. We have called $<$ an ordering, but the following synonyms would have done just as well; we could have called $<$ a *order* or a *partial order*. To add to the confusion, we may wish to refer to the ordered set $(\Omega, <)$ as the ordered set $(\Omega, >)$.

Example 8.2

Let Σ be a set. We denote the set of subsets of Σ by $P(\Sigma)$. The set $P(\Sigma)$ is called the power set of Σ (because if Σ is finite of cardinality n then $P(\Sigma)$ has cardinality 2^n). The binary operation $\subset$ (strict) is an ordering on $P(A)$. Notice that the second axiom of orderings prohibits $a < a$ for any $a \in \Omega$. Our axioms are designed to capture the (vague) notion of *strictly less than*. Some people prefer a slightly different set-up which captures the idea of *less than or equal to*. This issue is one of consuming unimportance, and the reader is invited to view the debate with fervent equanimity. The notation $a \leq b$, and similar variations, is used to express the proposition that either $a < b$ or $a = b$.

The example above (a power set ordered by inclusion) is quite a useful one to think about as a prototype for partial orders. It is true that it does have some special properties not enjoyed by arbitrary orders, but nonetheless it is a useful model, as long as you do remember that it has some features which make it special. For example, if α and β are distinct incomparable elements of $P(\Sigma)$ – so neither $\alpha \subset \beta$ nor $\beta \subset \alpha$ – then $\gamma = \alpha \cap \beta$ has the property that $\gamma \subset \alpha$ and $\gamma \subset \beta$. Moreover, whenever both $\delta \subset \alpha$ and $\delta \subset \beta$ then $\delta \subseteq \gamma$. For an arbitrary ordered set $\Omega, <$ if a, b are distinct incomparable elements of Ω we have no guarantee at all that there exists $c \in \Omega$ such that $c < a$ and $c < b$. To see this, let $\Omega = \{a, b\}$ and let $<$ be the empty relation. So, alas, the inclusion ordering on power sets is a bit special. However, it is not so special as the usual order on the natural numbers. This ordering on $\mathbb{N}$ has many useful properties not enjoyed by arbitrary orderings. For example, if x, y are integers then exactly one of the following statements is true: (a) $x < y$ (b) $y < x$ (c) $x = y$. We say (or rather other people do, so we had better join in) that the usual ordering $<$ satisfies *trichotomy*. In other words, exactly one of three things must happen.

Definition 8.3

If a partial order $<$ on a set Ω satisfies trichotomy then we say $<$ is a *total order* or (and this alternative is geometrically suggestive) $<$ is a *linear order*.

There are plenty of other attractive orderings on the natural numbers $\mathbb{N}$. For example, we may write $a < b$ as (bizarre) notation for the statement that $a \mid b$ but $a \neq b$. In words we might say that a is a proper divisor of b.

A widely used total order is the *lexicographic* ordering. This is the ordering used by the compilers of dictionaries. Let A denote the set of lower case letters in the roman alphabet, so $|A| = 26$ and $A = \{a, b, c, \ldots, z\}$. We form a new set A^* whose elements are finite *words* or *strings* in this alphabet. Thus $nibble \in A^*$ and $qwertyuiop \in A^*$. We even allow the empty word to be an element of A^*. In fact, we insist. The cleanest way to make precise the lexicographic ordering is to introduce a padding letter $\#$ to represent "no letter here folks", and adjoin a countable infinity of padding symbols to the end of each word. Thus cat becomes $cat\#\#\#\#\#\#\#\# \ldots$. The symbol $\#$ is invisible to everyone who is not a mathematician. To save invisible ink we write the infinite constant sequence of the symbol $\#$ as $\#^*$. Thus cat is written $cat\#^*$.

We can now think of our words as all being of infinite length, with only finitely many letters not being $\#$. Furthermore, only the invisible letter $\#$ can follow the letter $\#$. We really have an alphabet of 27 letters. We now describe the lexicographic ordering written $<_{lex}$. Write α for the word $\alpha_1\alpha_2\alpha_3 \ldots$. First order the letters as follows: $\# < a < b < \ldots$, this being the alphabetical order with the padding symbol thrown in at the front. Suppose β, γ are words then decide which is "larger" in the $<_{lex}$ ordering by finding the first i such that $\beta_i \neq \gamma_i$ and then comparing β_i with γ_i using our ordering on the alphabet. Thus $sequence <_{lex} sesquilinear$ because $q < s$. Moreover $caravan <_{lex} caravanserai$ since $\# < s$. Notice that there exists a first word in this ordering, the empty word $\#^*$. Although the empty word is not in English usage, lexicographers regard it as so important that they always include it in its correct position in the dictionary, and insert it right at the front just before the indefinite article. Check this in all available dictionaries.

Although dictionaries only include words of a language, the ordering $<_{lex}$ will serve to discriminate between all possible words. For example,

$$aaa <_{lex} aba <_{lex} ac <_{lex} ramanujan <_{lex} weierstrass.$$

We expand a little on the topic of lexicographic orderings. Take any non-empty finite alphabet Σ, select a total ordering of the letters, and form the lexicographic ordering as before. This ordering has a very useful property. Suppose $\omega_1, \omega_2, \nu \in \Sigma$ and $\omega_1 <_{lex} \omega_2$ then $\nu\omega_1 <_{lex} \nu\omega_2$. Thus the ordering is preserved under left-concatenation. This is not true on the right unless the alphabet consists of a single letter. Suppose σ_1, σ_2 are distinct letters of Σ, and that in the ordering of letters, $\sigma_1 < \sigma_2$. Let $\omega_1 = \sigma_1$ and $\omega_2 = \sigma_1\sigma_1$ so $\omega_1 < \omega_2$. Concatenate on the right with $\nu = \sigma_2$ then

$$\omega_1\nu = \sigma_1\sigma_2 > \sigma_1\sigma_1\sigma_2 = \omega_2\nu.$$

We say that $\Sigma, <_{lex}$ is a *left translation invariant* ordering.

We can tinker with the lexicographic ordering in many ways. For example, we can read all words backwards. This is the reverse lexicographic ordering denoted $<_{rlex}$ on words in the roman alphabet A. Notice that this ordering is not simply the opposite of the ordering $<_{lex}$. For example, *apple* $<_{lex}$ *pear* and *apple* $<_{rlex}$ *pear*. Nor is it the same ordering since *crystal* $<_{lex}$ *palace* but *palace* $<_{rlex}$ *crystal*.

Another thing we might do is to discriminate between words on the basis of their length, and then in the event of a tie, use lexicographic ordering to resolve the impasse. We call this the shortlex or length-then-lexicographic ordering $<_{llex}$ on A^*. In this ordering we have *tiny* $<_{llex}$ *small* but of course *small* $<_{lex}$ *tiny*. Note that shortlex is a two-sided translation invariant total ordering.

We can generalize this. We can ascribe to each letter a *weight*, usually selected from the natural numbers. We can then define the weight of a word to be the sum of the weights of its constituent letters. Then word weight can be used as a primary discriminator. Some other criterion, say lexicographic ordering, can be used to break ties.

EXERCISES

Consider the following sentence: *Frank Clements was in Trafalgar Square at noon on January the first in the year two thousand.* Let the words of this sentence form a set X.

8.5 Put the elements of X into lexicographical order.

8.6 Put the elements of X into lexicographical order with the usual order of letters reversed ($z < y < \cdots < a$).

8.7 Put the elements of X into *llex* order.

8.2 Chain Conditions

The usual ordering on the natural numbers $\mathbb{N}$ has the property that if $S \subset \mathbb{N}$ and $S \neq \emptyset$, then there exists a minimal element s of S. This is an element s such that there is no $x \in S$ with $x < s$. The same guarantee cannot be offered regarding the existence of maximal elements; in the case where $S = \mathbb{N}$, no matter which $t \in S$ we select, there will be $x \in S$ such that $t < x$.

Definition 8.4

Suppose that Ω is a partially ordered set with ordering denoted by $<$, and that $S \subseteq \Omega$ and S is non-empty. We say that $\omega \in S$ is *minimal* in S if there does not exist $\nu \in S$ such that $\nu < \omega$.

We leave the reader to make a formal definition of a maximal element. As we have observed, maximal and minimal elements of a given subset in a partial ordering may not exist, and even if, say, a maximal element exists, it may not be unique.

Example 8.5

Let Π denote the set whose elements are the proper subsets of $\{1, 2, 3\}$. In this context, we mean to exclude both $\{1, 2, 3\}$ and $\emptyset$ from membership of Π. Thus $|\Pi| = 6$, and Π is ordered by subset inclusion. Now Π has 3 maximal and 3 minimal elements.

The situation becomes under control in the event that our ordering is total (linear).

EXERCISES

8.8 If Ω equipped with $<$ is a linearly ordered set, and $S \subseteq \Omega$ happens to contain a minimal element s then s is the unique minimal element of S.

Of course, we could replace the word minimal by the word maximal in that exercise.

Definition 8.6

Suppose that $(\Omega, <)$ is a partially ordered set with the property that if $S \subseteq \Omega$ is non-empty, then S contains at least one minimal element. The ordering $<$ is said to be *well-founded*.

Naturally any ordering on a finite set is well-founded, and as we have observed, the usual ordering on $\mathbb{N}$ is well-founded. The usual ordering on the integers $\mathbb{Z}$ is not well-founded (why not?). Suppose that $(\Omega, <)$ is any partially ordered set. We say that a subset C of Ω is a *chain* if the induced ordering on C is total. We say that the ordering satisfies the *descending chain condition* if there is no infinite chain

$$C_1 > C_2 > C_3 > \dots$$

where $C = \{C_i \,|\, i \in \mathbb{N}\}$. Similarly we can define the ascending chain condition. Historically mathematicians have called an ordering satisfying the ascending chain condition *Noetherian* (for the female algebraist Emmy Noether). Conversely orderings satisfying the descending chain condition have been called *Artinian* (for the male algebraist Emile Artin). The rewriting community seem to have partially reversed this convention, and refer to orderings satisfying the descending chain condition as Noetherian. We will dodge the issue by banning these eponymous adjectives.

The following classical theorem serves to explain the connection between well-foundedness and the descending chain condition:

Theorem 8.7

The following conditions on a partially ordered set Ω are equivalent:

(i) The ordering satisfies the descending chain condition.

(ii) If S is any non-empty subset of Ω then S contains a minimal element (i.e. the ordering is well-founded).

Proof

Suppose (i) holds, and (for contradiction) that there exists a non-empty subset S of Ω which contains no minimal element. Choose $s_1 \in S$ arbitrarily, and obtain an infinite descending chain of elements (s_i) inductively. Suppose we have defined s_i for all $i < j \in \mathbb{N}$. The element s_{j-1} cannot be minimal, so choose s_j to be any element of S such that $s_j < s_{j-1}$. The chain (s_i) violates condition (i) and we are half way home.

Now suppose condition (ii) holds. Suppose (for contradiction) that condition (i) failed to hold for the descending chain (s_i). Let $S = \{s_i \,|\, i \in \mathbb{N}\}$ then S contains no minimal element and so condition (ii) is violated.

$\square$

Let us return to the lexicographic ordering on the words in the lower case Roman alphabet. This ordering is not well-founded. The following infinite descending chain demonstrates this fact:

$$b >_{lex} ab >_{lex} aab >_{lex} aaab >_{lex} \dots$$

Thus it is fortunate that lexicographers only list legal words in the dictionary. Otherwise there would be infinitely many entries before one reached the section of words beginning with the letter b. This ordering does not satisfy the ascending chain condition either. Notice that

$$a < aa < aaa < \ldots$$

In this ordering on words we have:

$$zoo < zoigma < aardvark < antelope < abbreviation < asdfghjklzxcvbnm.$$

We have obtained a well-founded order.

8.3 Zorn's Lemma

The name Zorn's Lemma is (from some points of view) a misnomer. Many working algebraists are quite happy to view Zorn's Lemma as an axiom. As such it does not require a proof and is therefore not a lemma. Let us get the statement out of the way.

Lemma 8.8 (Zorn)

Suppose $(\Omega, <)$ is a partially ordered set and $S \subseteq \Omega$. Suppose furthermore that whenever C is a chain in S then there exists $c_0 \in S$ such that $c < c_0$ for each $c \in C$. We allow ourselves to deduce that there exists an element $c_* \in S$ which is maximal in S.

Ponder on this for a while. It may seem blindingly obvious to you that Zorn's lemma is "true". Alternatively, it may give off the aroma of rotting fish. It turns out to be equivalent to the *Axiom of Choice*. This is the axiom that allows you to pick a transversal for any set of non-empty sets. An axiom is necessary since the usual foundations of set theory allow you to pick an element of a non-empty set, but to do this simultaneously for an infinite collection of sets is a problem. If the collection of sets is countable then one can use a recursive procedure, but that is not a solution for the general case. Our collections of sets may well be uncountable, and we need a handle on such vast aggregations.

Now, given one of the usual (fashionable) foundations of mathematics, say Zermelo–Fraenkel Set Theory (ZF), Zorn's Lemma and the Axiom of Choice are equivalent. Assume one, and the other comes for free. This does not make either of them *true* (whatever that means). There are perfectly sane people who do interesting mathematics who are not prepared to take either principle on board. Nonetheless, the vast majority of mathematicians do accept both as a sensible device for reasoning about the infinite. Of course, fashions may change. Zermelo–Fraenkel set theory may go the way of the dodo, polywater and phlogiston.

Within the context of ZF, there is yet another equivalent formulation of Zorn's Lemma/the Axiom of Choice. This is the *Principle of Well-Ordering*. This asserts that given any set Ω, there exists a well-founded total order on Ω. In case you think that this is so very reasonable that we should pass this principle without a fuss, think about applying this principle to the real numbers.

The authors privately regard the Axiom of Choice as being very plausible, Zorn's Lemma as being a necessary convenience, and the Principle of Well Ordering as being preposterous. Nonetheless, they are equivalent (so our confession is just an insight into our tortured minds). Let us illustrate Zorn's Lemma in action.

Theorem 8.9

Every vector space V has a basis.

Proof

Order the linearly independent subsets of V by inclusion. Let C be any (ascending) chain in our ordering. We aim to find an upper bound for C. This is easy. Let $U = \cup_{c \in C} c$ then U is a linearly independent set (why?). Thus we may apply Zorn's Lemma to conclude that V contains a subset maximal with respect to the property of being linearly independent. Call this set M. Now look at $\langle M \rangle$, the span of M. If $\langle M \rangle \neq V$ then choose $x \in V - \langle M \rangle$. The set $M_1 = M \cup \{x\}$ must be linearly independent (else we force $x \in \langle M \rangle$). This contradicts the maximality of the linearly independent set M. Thus $\langle M \rangle = V$ and M is a basis of V.

$\square$

With Zorn's Lemma in the tool box, the infinite looks a deal less frightening. Zorn's lemma applies in a descending context of course. We can either adopt the corresponding descending/minimal version of the lemma, or sidestep the issue as follows; given an ordered set $(\Omega, <)$ we can define the opposite ordered set $(\Omega, <_{op})$. This ordering is defined by $a <_{op} b$ exactly when $b < a$. Zorn's lemma applied to $<_{op}$ then translates into the descending version for $<$.

For well-founded orderings, descending chains are finite and minimal elements of non-empty subsets are guaranteed anyway.

Further Reading

There are many excellent abstract algebra books including the magnificent *Topics in Algebra* by I.N. Herstein (John Wiley, 1975) and the more recent (and also splendid) *Contemporary Abstract Algebra* by J.A. Gallian (Houghton Mifflin, 1998). We also draw your attention to *A Survey of Modern Algebra, 1997* by G. Birkhoff and S. Mac Lane (A K Peters Ltd) and *Algebra* by the conveniently transposed S. Mac Lane and G. Birkhoff (New York: Macmillan 1967). P.M. Cohn's *Algebra* I, II and III (John Wiley, 1982) is very comprehensive as is the third edition of *Algebra* by S. Lang (Addison Wesley, 1993).

As for group theory, an excellent text at about the level of this book is J.E. Humphreys's *A Course in Group Theory* (O.U.P. 1996). Our chapter 5 was informed by Philip Hall's influential QMC (now QMW) notes and D. Segal's highly regarded *Polycyclic Groups* (C.U.P. 1983).

Introduction to the Theory of Groups by J.J. Rotman (Springer, 1995) would be a good next text to read, and *A Course in the Theory of Groups* by D.J.S. Robinson (Springer 1982) is the happy group theorist's bed-time companion.

The subject of group presentations was introduced in Chapter 6. We commend *Presentations of Groups* by D.L. Johnson (C.U.P. 1997) as a very good starting point for this subject. After that the classical *Combinatorial Group Theory* by W. Magnus, A. Karrass, D. Solitar (New York 1966 (also Dover)) makes good reading, as does *Combinatorial Group Theory* by R.C. Lyndon and P.E. Schupp (Springer, 1967). For a more recent text concerning connections between presentation theory, geometry and formal language theory, see *Word Processing in Groups* by D.B.A. Epstein et al. (Jones and Bartlett, 1992).

We also make passing reference to representation theory. The charming and beautifully constructed *Representations and Characters of Groups* by G. James and M. Liebeck (C.U.P. 1993) is both inspiring and accessible.

Solutions

Chapter 1

1.1 (ii) Postmultiply the equation $y * x = z * x$ by x' which exists by the Inverse Axiom. Therefore $(y * x) * x' = (z * x) * x'$. Now by the Associative Axiom we have $y * (x * x') = z * (x * x')$ and so $y * e = z * e$. However e is an identity element so $y = z$.

(iii) Let $z = x' * y$ so that $x * z = x * (x' * y) = (x * x') * y$ by the Associative Axiom. Now $x * x' = e$ by the Inverse Axiom so $x * z = e * y = y$ as required.

(iv) Let $w = y * x'$, then $w * x = (y * x') * x = y * (x' * x)$ by the Associative Axiom. Now $x' * x = e$ by the Inverse Axiom so $w * x = y * e = y$ as required.

(v) Suppose that $x * y = x * e$, and deploy the result proved in part (i).

(vi) Suppose that $y * x = e * x$, and deploy the result proved in part (ii).

1.2 Certainly $e * e = e$. Let x be the non-identity element of G, then if $x * x = x$ then $x = e$ by cancellation (Lemma 1.1), which is absurd. Thus $x * x = e$.

1.3 Certainly $e * e * e = e$. Let $G = \{e, x, y\}$. If $x * y = x$ or $x * y = y$ then $x = e$ or $y = e$ respectively (which are absurdities), so $x * y = e$. Now if $x * x = e$ we have $x = y$ by uniqueness of inverses which is also absurd. If $x * x = x = e * x$, then $x = e$ by cancellation, which is nonsense. Therefore $x * x = y$ so $x * x * x = x * (x * x) = x * y = e$. The problem is symmetric in x and y so $y * y * y = e$.

1.4 This operation is closed and associative. Every element is an identity on the right, but no element is an identity on the left. If you choose a right identity θ then each letter ψ has the property that $\theta \psi = \theta$, and so there are left inverses with respect to any chosen right identity.

1.5 Certainly if $x, y \in G$ then $x \Box y = y * x \in G$. Thus closure is established. The element 1 is a two sided identity for $\Box$ and the $\Box$-inverse of $x \in G$ is x^{-1}. If $x, y, z \in G$ then

$$
\begin{aligned}
(x \Box y) \Box z &= z(x \Box y) = z * (y * x) = (z * y) * x \\
&= (y \Box z) * x = x \Box (y \Box z).
\end{aligned}
$$

1.6 Using the axioms one at a time we have

$$
x' * x * x' * (x')' = (x' * x) * (x' * (x')') = (x' * x) * e = x' * x
$$

but also

$$x' * x * x' * (x')' = (x' * (x * x')) * (x')' = (x' * e) * (x')' = x' * (x')' = e$$

so $x' * x = e$ for every $x \in G$ so left inverses exist and coincide with right inverses. It remains to show that e is a left identity. Suppose that $x \in G$. It follows that $e * x = (x * x') * x = x * (x' * x) = x * e = x$ so e is a left identity and G is a group.

1.7 Suppose that $x^{-1} = y^{-1}$, then $(x^{-1})^{-1} = (y^{-1})^{-1}$ so by Corollary 1.2 we have $x = y$. Thus inversion is injective. Now if $x \in G$, then $x = (x^{-1})^{-1}$ as we have just observed, so x is the inverse of x^{-1} and therefore inversion is surjective.

1.8 (a) $1 \in A \neq \emptyset$. Now suppose that $a, b \in A$, so $a^n = b^n = 1$. Now $(ab^{-1})^n = a^n(b^n)^{-1} = 1$ so $ab^{-1} \in A$ so $A \leq G$.

   (b) $1 = 1^n \in B$ so $B \neq \emptyset$, Suppose that $a, b \in B$, then $a = c^n$ and $b = d^n$ for some $c, d \in B$. Now $(cd^{-1})^n = c^n(d^n)^{-1} = ab^{-1}$ so $B \leq G$.

   (c) $1 = 1^1 \in C \neq \emptyset$. Now suppose that $a, b \in C$ so there natural numbers u, v such that $a^u = 1 = b^v$. Now

$$(ab^{-1})^{uv} = (a^u)^v \cdot ((b^v)^u)^{-1} = 1 \cdot 1 = 1$$

   so $C \leq G$.

1.9 $1 \in H$ so $1 = g^{-1}1g \in D \neq \emptyset$. Suppose that $x, y \in g^{-1}Hg$, so there are $h, k \in G$ with $x = g^{-1}hg$ and $y = g^{-1}kg$. Now $y^{-1} = g^{-1}k^{-1}g$ so $xy^{-1} = g^{-1}hg \cdot g^{-1}k^{-1}g = g^{-1}hk^{-1}g \in D$. Thus $D \leq G$.

1.10 Since the family is non-empty, we have $1 \in H_i$ for at least one (and in fact all) $i \in I$ so $1 \in \cup_{i \in I} H_i \neq \emptyset$. Now suppose that $x, y \in \cup_{i \in I} H_i$. Thus there are $m, n \in I$ such that $x \in H_m$ and $y \in H_n$. Without loss of generality $H_m \leq H_n$, so $x, y \in H_n$. Now $xy^{-1} \in H_n$ so $xy^{-1} \in \cup_{i \in I} H_i$. Therefore $\cup_{i \in I} H_i \leq G$.

1.11 Let $x \mapsto x$ for $x < 0$ and $x \mapsto x + 1$ for $x \geq 0$.

1.12 (a) Two maps were mentioned in the example: $q \mapsto q + 1$ and $q \mapsto 2q$. Call these maps α, β respectively. Now $(0)(\alpha \circ \beta) = 2$ and $(0)(\beta \circ \alpha) = 1$ so $\alpha \circ \beta \neq \beta \circ \alpha$ and therefore A is not an abelian group.

   (b) It suffices to prove the result for either γ or γ^{-1}. Since γ is not the identity map, there is $q \in \mathbb{Q}$ which is not fixed by γ. Either $q < (q)\gamma$ or $(q)\gamma < q$. In the latter event $q < (q)\gamma^{-1}$. Replacing γ by γ^{-1} if necessary, we may assume that $q < (q)\gamma$. Now γ respects the order so by an induction argument we have $(q)\gamma^n < (q)\gamma^{n+1} \ \forall n \in \mathbb{N}$. The case of negative exponents can also be dealt with by induction so $(q)\gamma^n < (q)\gamma^{n+1} \ \forall n \in \mathbb{Z}$. We conclude that if $i, j \in \mathbb{Z}$ and $i \neq j$, then $(n)\gamma^i \neq (n)\gamma^j$, so $\gamma^i \neq \gamma^j$.

1.13 Suppose that $AB \leq G$. If $a \in A$ and $b \in B$, then $ba = (a^{-1}b^{-1})^{-1} \in AB$ since A, B and AB are groups. Therefore $BA \subseteq AB$. Also $ab = (a'b')^{-1}$ for some $a' \in A, b' \in B$. Therefore $ab = b'^{-1}a'^{-1} \in BA$. Thus $AB \subseteq BA$. We conclude that $AB = BA$. Conversely suppose that $AB = BA$. Now AB is non-empty by design, and $(AB)(AB) = (A(BA))B = (AA)(BB) = AB$ so AB is multiplicatively closed. Finally if $a \in A, b \in B$ then

$$(ab)^{-1} = b^{-1}a^{-1} \in BA = AB$$

so $AB \leq G$.

1.14 Suppose that $H \not\subseteq K$, so there is $h \in H - K$. Choose any $k \in K$, and then hk is the product of two elements of $H \cup K$. We suppose this to be a group so $hk \in H \cup K$. Now $h \notin K$ so $hk \notin K$, and therefore $hk \in H$. It follows that $k \in H$. Therefore $K \subseteq H$. In either event $H \cap K \in \{H, K\}$. Now for the reverse implication. Suppose that $H \cap K \in \{H, K\}$. Now $H \cap K$ is a subgroup of both H and K, so either $H \leq K$ or $K \leq H$. In the first event $\langle H, K \rangle = K = H \cup K$. In the second case $\langle H, K \rangle = H = H \cup K$.

1.15 Use the Klein Vierergruppe, with H, K, L the distinct subgroups of order 2.

1.16 Suppose that $H = \langle X \rangle$ for some finite set X. Now $|G : H| < \infty$ so there is a finite set Y such that $G = HY$. Now every element of G is expressible as hy for $h \in H, y \in Y$, and so as a word on letters of $X \cup Y$ and their inverses. Therefore $G = \langle X, Y \rangle = \langle X \cup Y \rangle$ is finitely generated.

1.17 Let Z be a finite generating set for G. Express each $z \in Z$ as a word on the generating set Y. Form the finite set X which consists of the elements of Y which are mentioned in at least one of these finitely many words. Now every element of G is expressible as a word on the elements of Y (and their inverses), and every element of Y is expressible as a word on the elements of X (and their inverses). Therefore every element of G is expressible as a word on the elements of X (and their inverses) so $G = \langle X \rangle$.

1.18 Every element of D which is not in Z has the form $z^n x$ for some $n \in \mathbb{Z}$. This follows from the theory of dihedral groups which we have developed. If $n = 0$ there is nothing to prove since $x^2 = 1$. Assume that $n > 0$ and induct on n. Now

$$z^n x z^n x = z(z^{n-1}x)(z^{n-1}x)xzx = zxzx = xyxxyx = 1.$$

Assume that $n < 0$ and put $m = -n$. We have

$$((z^n x)^2)^{-1} = xz^m xz^m = x(z^m x)(z^m x)x = x^2 = 1$$

so $(z^n x)^2 = 1$. Finally we had best be certain that $z^n x \neq 1$, but this cannot happen as it would entail $x \in \langle z \rangle$ and the theory has eliminated this possibility.

1.19 The only way that x, y can fail to generate D_{2p} is if they both inhabit the same proper subgroup. Recall that C_p has index 2 in D_{2p}. We count pairs (a, b) which do not generate $D = D_{2p}$. Recall that C_p has index 2 in D_{2p}. We can have $a = 1, b \in D$ ($2p$ possibilities) or $a \in C_p$ but $a \neq 1$, $b \in C_p$ ($(p-1)p$ possibilities). Finally we may have $a \in D - C_p$ and $b = 1$ or $a \in D - C_p$ and $ab = 1$ ($2p$ possibilities). The probability that a, b do not generate D is therefore

$$(2p + (p-1)p + 2p)/(2p)^2 = p^2 + 3p/4p^2 = 1/4 + 3/4p.$$

The probability that $\langle a, b \rangle = D$ is therefore $3(1 - 1/p)/4$.

1.20 We count pairs (a, b) which do not generate $D = D_{2p^t}$. Recall that C_{p^t} has index 2 in D_{2p^t}. We can have $a \in C_{p^{t-1}}, b \in D$ ($2p^{2t-1}$ possibilities) or $a \in C_{p^t} - C_{p^{t-1}}$, $b \in C_{p^t}$ ($(p^t - p^{t-1})p^t = p^{2t} - p^{2t-1}$ possibilities). Finally we may have $a \in D - C_{p^t}$ and $b \in C_{p^{t-1}}$ or $a \in D - C_p$ and $ab \in C_{p^{t-1}}$ ($2p^{t-1}$ possibilities). The probability that a, b do not generate D is therefore

$$(2p^{2t-1} + p^{2t} - p^{2t-1} + 2p^{t-1})/4p^{2t} = 1/4 + 3/4p.$$

The probability that $\langle a, b \rangle = D$ is therefore $3(1-1/p)/4$. This answer is therefore independent of t.

Chapter 2

2.1 Suppose that $x, y \in H$, then $\varphi|_H : xy \mapsto (xy)\varphi = (x)\varphi \cdot (y)\varphi$. However, $(x)\varphi|_H \cdot (y)\varphi|_H = (x)\varphi \cdot (y)\varphi$ so $\varphi|_H$ is a homomorphism of groups.

2.2 Proposition 2.2(i) shows that the result holds when $t = 0$. There is nothing to prove when $t = 1$ and for $t > 1$ we proceed by induction. We have $(x^t)\alpha = (x^{t-1}x)\alpha = (x^{t-1})\alpha \cdot (x)\alpha = ((x)\alpha)^{t-1} \cdot (x)\alpha = ((x)\alpha)^t$. Next suppose that $t < 0$. Now $(g^t)\alpha = ((g^{-t})^{-1})\alpha = ((g^{-t})\alpha)^{-1} = (((g)\alpha)^{-t})^{-1} = (g)\alpha^t$.

2.3 $(x^{-1}yx)\beta = (x^{-1}y)\beta \cdot (x)\beta = ((x)\beta)^{-1} \cdot (y)\beta) \cdot (x)\beta = ((y)\beta \cdot ((x)\beta)^{-1}) \cdot (x)\beta$
$= (y)\beta(((x)\beta)^{-1} \cdot (x)\beta) = (y)\beta \cdot 1 = (y)\beta$.

2.4 Suppose that $h, k \in H$. Now γ is surjective so there are $x, y \in G$ such that $(x)\gamma = h$ and $(y)\gamma = k$. Therefore $hk = (x)\gamma \cdot (y)\gamma = (xy)\gamma = (yx)\gamma$ since G is abelian. Therefore $hk = (y)\gamma \cdot (x)\gamma = kh$ so H is abelian.

2.5 We have $((g)\delta)^n = (g^n)\delta = (1)\delta = 1$. Thus $(g)\delta$ has finite order m. Now the fact that $(g)\delta^n = 1$ forces $m \mid n$ by Remark 1.42.

2.6 Let $H = \{h \mid h \in G, (h)\theta_1 = (h)\theta_2\}$ it is easy to verify that H is a subgroup of G and $X \subseteq H \leq G$ so $\langle X \rangle \leq H$ since $\langle X \rangle$ is the smallest subgroup of G which contains X. Therefore $\langle X \rangle \leq H \leq G = \langle X \rangle$. Thus $H = G$. Alternatively you can argue that every element is expressible as a word on elements from X an their inverses, and exploit the fact that θ is a homomorphism.

2.7 $(ab)^2 = 1$ so $abab = 1$ so $ab = baabab = ba$ for all $a, b \in G$.

2.8 By Proposition 2.11 it follows that $MN \trianglelefteq G$. The remaining issue is normality. Suppose that $m \in M, n \in N$ and $g \in G$.b Now $(mn)^g = m^g n^g \in MN$ since $m^g \in M$ and $n^g \in N$ since $M, N \trianglelefteq G$. Now Proposition 2.8 applies.

2.9 Suppose that $n \in N$, so $n \in M_i$ for each $i \in I$, so if $g \in G$, then $n^g \in g^{-1}M_i g = M_i$ for each $i \in I$. Thus $g^{-1}ng \in N$ for every $g \in G, n \in N$ so $N \trianglelefteq G$ by Proposition 2.8.

2.10 Clearly $H \cap N \leq H$. Suppose that $x \in H \cap N$ and $h \in H$. Now $x^h \in H$ since $x \in H$, and $x^h \in N$ since $N \trianglelefteq G$. Therefore $x^h \in H \cap N$. However x and h were arbitrary so $H \cap N \trianglelefteq H$.

2.11

$$\begin{pmatrix} 2 & 0 \\ 0 & 1 \end{pmatrix} \begin{pmatrix} 1 & x \\ 0 & 1 \end{pmatrix} \begin{pmatrix} 1/2 & 0 \\ 0 & 1 \end{pmatrix} = \begin{pmatrix} 1 & 2x \\ 0 & 1 \end{pmatrix}$$

Let H be the set of all matrices of the form

$$\begin{pmatrix} 1 & x \\ 0 & 1 \end{pmatrix}$$

where $x \in \mathbb{Z}$. Let

$$A = \begin{pmatrix} 1/2 & 0 \\ 0 & 1 \end{pmatrix}$$

and let $G = \langle \{A\} \cup H \rangle$ be a subgroup of the set of invertible 2×2 matrices with rational entries. Now H^A consists of those elements of H with top right entry an even number.

2.12 There is no doubt that H is a proper subgroup, so normality is the only issue. If $g \in H$ then $gH = H = Hg$. If $g \notin H$, then $gH \neq H \neq Hg$. Since $|G : H| = 2$ there is only one left (or right) coset other than H, and it must be $G - H$. Therefore $gH = G - H = Hg$. We are done.

2.13 Suppose that $G/H = \langle tH \rangle$. Now $G = \langle t, H \rangle$ and since all elements of $\{t\} \cup H$ commute pairwise, it follows that G is abelian.

2.14 Yes. They are both isomorphic to C_{1729} by the structure theorem for finite abelian groups.

2.15 This is the number $p_2(n)$ of partitions of n into at most two parts. Proof by induction on n (doing odd and even separately).

2.16 This is the number $p_3(n)$ of partitions of n into at most three parts. It is helpful to do 6 base cases $n = 0, \ldots, 5$ and then for $n \geq 6$ to use the recurrence $p_3(n) = p_2(n) + p_2(n-3) + p_3(n-6)$ because it is not hard to show that $p_2(n) + p_2(n-3) = n$. Now use induction on n in each of the 6 cases depending on the congruence class of $n \bmod 6$.

2.17 The maximum order of an element in S_3 is 3, but there are also elements of order 2. This constitutes a bizarre proof that S_3 is not abelian.

Chapter 3

3.1 (a) This is not a group since it does not contain the identity permutation.
   (b) This is a group. It contains the identity element. Take the argument which shows that the finitary permutations form a group and adjust it for this purpose. Observing that if $\sigma_1, \sigma_2 \in$ Sym Ω, then $\operatorname{supp}(\sigma_1\sigma_2) \subseteq \operatorname{supp}(\sigma_1) \cup \operatorname{supp}(\sigma_2)$ and that the union of two countable sets is countable.
   (c) There are no permutations with support of size 1, so the set described is the trivial permutation group on Ω, and that is a group.
   (d) The given set is not closed under composition provided Ω has at least 3 distinct elements ω_1, ω_2 and ω_3. Consider the permutation α which has support $\{\omega_1, \omega_2\}$ and β which has support $\{\omega_2, \omega_3\}$. It is easy to verify that $\alpha\beta$ has support $\{\omega_1, \omega_2, \omega_3\}$.
3.2 Let $\Omega_1 = \{1, 2, \ldots, n\}$ and $\Omega_2 = \{n+1, \ldots, 2n\}$ and put $\Omega = \Omega_1 \cup \Omega_2$. Now put $A = $ Sym Ω_1, $B = $ Sym Ω_2. Now let $H = \langle A, B \rangle$. Each of A and B is normal in H, so $H = AB$. Moreover $A \cap B = 1$ by design. Now Definition 2.37 applies, and $H = A \times B \simeq S_n \times S_n$ is a direct product.
3.3 Let the cycles be $\eta_1, \ldots, \eta_t$. These cycles have respective orders (lengths) $n_1, \ldots, n_t$ and commute with one another. Now $\sigma^m = 1$ if and only if $\eta_i^m = 1$ for every i since the supports of the η_j are pairwise disjoint. Therefore $\sigma^m = 1$ if and only if m is a common multiple of $n_1, \ldots, n_t$. The order of σ is therefore the least common multiple of $n_1, \ldots, n_t$.
3.4 We may assume that $\Omega = \{1, 2, 3, \ldots, 13\}$ so Sym $\Omega = S_{13}$. The cycles $(1, 2, \ldots, m)$ have order m so all orders up to 13 arise. Now

$$(1, 2, 3, 4, 5, 6, 7)(8, 9)$$

has order 14, whereas the permutation

$$(1, 2, 3, 4, 5)(6, 7, 8)$$

has order 15. There is no element of S_{13} which has order 16 because 16 is not the least common multiple of natural numbers smaller than itself, since it is a power of a prime number. See Exercise 3.3.
3.5 There are many ways to do this. Here is one. Suppose $|z|$ is not a prime power, then $z \mapsto z$. From now on we assume that $|z| = p^i$ for some prime number p. We map such elements as shown:

$$\cdots \mapsto -p^2 \mapsto -p \mapsto p \mapsto p^2 \mapsto p^3 \mapsto \cdots$$

and $1 \mapsto 1$. There are infinitely many prime numbers (according to Euclid) so this permutation involves infinitely many disjoint infinite cycles.
3.6 Let G be the group of symmetries of the uncoloured flag. Since the flag will be coloured the same on the back as the front, we may take G to be cyclic of order 2, where the non-identity element rotates the flag through π about its centre. Burnside's formula yields that the number of essentially different Eurostandard standards is $(c^3 + c^2)/2$.
3.7 The cube: the 24 rigid symmetries of a cube come in various sorts (conjugacy classes in fact). There is the identity. There are rotations through π about an axis joining the centre of an edge to the centre of the antipodal edge. There are

rotations through π about an axis joining the centre of a face to the centre of the opposite face, and rotations through $\pm\pi/2$ about a similar axis. There are four grand diagonals; straight line segments joining a vertex to its antipode. Rotations through $\pm 2\pi/3$ are rigid symmetries.

element type	No. of els.	$|\text{Fix}_V|$	$|\text{Fix}_E|$	$|\text{Fix}_F|$
identity	1	c^8	c^{12}	c^6
ce-ce rot by π	6	c^4	c^7	c^3
cf-cf rot by $\pm\pi/2$	6	c^2	c^3	c^3
cf-cf rot by π	3	c^4	c^6	c^4
grd diag rot by $\pm 2\pi/3$	8	c^4	c^4	c^2

You can count the number of distinct colourings using the following three polynomials:

The vertex colouring polynomial: $(c^8 + 17c^4 + 6c^2)/24$
The edge colouring polynomial: $(c^{12} + 6c^7 + 3c^6 + 8c^4 + 6c^3)/24$
The face colouring polynomial: $(c^6 + 3c^4 + 12c^3 + 8c^2)/24$

The regular tetrahedron:

element type	No. of els.	$|\text{Fix}_V|$	$|\text{Fix}_E|$	$|\text{Fix}_F|$
identity	1	c^4	c^6	c^4
vx-cf rot by $\pm 2\pi/3$	8	c^2	c^2	c^2
ce-ce rot by π	3	c^2	c^4	c^2

You can count the number of distinct colourings using the following three polynomials:

The vertex colouring polynomial: $(c^4 + 11c^2)/12$
The edge colouring polynomial: $(c^6 + 3c^4 + 8c^2)/12$
The face colouring polynomial: $(c^4 + 11c^2)/12$

The regular dodecahedron: this figure has 12 faces, each of which is a regular pentagon of identical size. It has 30 edges and 20 vertices.

element type	No. of els.	$|\text{Fix}_V|$	$|\text{Fix}_E|$	$|\text{Fix}_F|$
identity	1	c^{20}	c^{30}	c^{12}
vx-vx rot by $\pm 2\pi/3$	20	c^8	c^{10}	c^4
ce-ce rot by π	15	c^{10}	c^{16}	c^6
cf-cf rot by $\pm 2\pi/5, \pm 4\pi/5$	24	c^4	c^6	c^4

You can count the number of distinct colourings using the following three polynomials:

The vertex colouring polynomial: $(c^{20} + 15c^{10} + 20c^8 + 24c^4)/60$
The edge colouring polynomial: $(c^{30} + 15c^{16} + 20c^{10} + 24c^6)/60$
The face colouring polynomial: $(c^{12} + 15c^6 + 44c^4)/60$

3.8 (a) Let $\Gamma = \{(x, y) \mid x, y \in G, xy = yx\}$. Thus the required probability is

$$|\Gamma|/|G|^2 = \frac{1}{|G|^2} \sum_{x \in G} |C_G(x)| = t/|G|$$

where t is the number of conjugacy classes of G. Note the crucial use of not Burnside's counting principle. When G is abelian, this gives a probability of 1, which is correct. When $G \simeq S_3$ then an inspection of the multiplication table yields that 18 of the 36 entries coincide with the element in the transposed position. Now $18/36 = 1/2 = 3/6$, as given by the formula.

(b) If x is fixed then the number of conjugates of x is $|G : C_G(x)|$. Thus the probability that a (uniformly randomly chosen) x and y are in the same conjugacy class is $|\Lambda|/|G|^2$ where

$$\Lambda = \{(x, y) \mid x, y \in G, \exists t \in G \text{ s.t. } x^t = y\}.$$

Now $|\Lambda| = \sum_{x \in G} |G : C_G(x)|$ so the answer is

$$\frac{1}{|G|} \sum_{x \in G} \frac{1}{|C_G(x)|}.$$

Thus the probability that two uniformly randomly chosen elements are conjugate is the average size of the reciprocal of the centralizer of an element. Note that this answer is correct if G is abelian. Incidentally, this means that the probability that a randomly chosen pair of elements fo S_3 are conjugate is $7/18$.

3.9 (a) Let t denote the number of the G-conjugacy classes which are contained in N. Now G acts on N by conjugation. Now

$$t = \frac{1}{|G|} \sum_{g \in G} |C_N(g)|.$$

Therefore

$$nt = \frac{1}{|G|} \sum_{g \in G} |G : N||C_N(g)|.$$

Now $|C_G(g) : C_N(g)| = |C_G(g) : N \cap C_G(g)|$ and

$$C_G(g)/N \cap C_G(g) \simeq C_G(g)N/N \leq G/N.$$

Therefore $|C_G(g) : C_N(g)| \leq n = |G : N|$. We deduce that

$$nt \geq \frac{1}{|G|} \sum_{g \in G} |C_G(g) : C_N(g)||C_N(g)| = \frac{1}{|G|} \sum_{g \in G} |C_G(g)|.$$

However, this last expression is the number of conjugacy classes of G. Divide through by n and we are done.

(b) Let Γ denote the union of the G-conjugacy classes which intersect H non-trivially, and suppose that there are t such G-conjugacy classes. Now G acts on Γ by conjugation and

$$t = \frac{1}{|G|} \sum_{g \in G} |C_\Gamma(g)|$$

and $C_\Gamma(g)$ is the set of elements of Γ which commute with g. Now $C_H(g) \subseteq C_\Gamma(g)$ for each $g \in G$ so

$$t \geq \frac{1}{|G|} \sum_{g \in G} |C_\Gamma(g)|.$$

Now

$$nt \geq \frac{1}{|G|} \sum_{g \in G} |G : H||C_H(g)|$$

This time we do not have the isomorphism theorems to hand because H is not necessarily normal in G. However

$$|C_G(g) : C_H(g)| = |H \cap C_G(g) \backslash C_G(g)| = |H \backslash C_G(g)H|$$

by Proposition 2.30 and so $|C_G(g) : C_H(g)| \leq |G : H| = n$ since $C_G(H)H \subseteq G$. Now

$$nt \geq \frac{1}{|G|} \sum_{g \in G} |C_G(g)|$$

and we finish by dividing by n.

(c) The first result is simply a special case of the second. When N is normal in G, if a G-conjugacy class intersects N non-trivially, then it must be contained in N.

3.10 Suppose that G acts on Ω. We define $\beta : G \to \text{Sym } \Omega$ via $(g)\beta : \omega \mapsto \omega \cdot g$ for every $\omega \in \Omega$ and every $g \in G$. Now we use β to define an action on Ω as follows: $\omega * g = (\omega)((g)\beta) = \omega \cdot g$ for every $\omega \in \Omega$ and $g \in G$. Therefore the two actions coincide.

Start again, this time with a homomorphism $\beta : G \to \text{Sym } \Omega$. Define an action $*$ of G on Ω as above. Now the action $*$ induces a homomorphism $\gamma : G \, rightarrow \text{Sym } \Omega$ via $(g)\gamma : \omega \mapsto \omega * g$ for every $g \in G$ and every $\omega \in \Omega$. However, $\omega * g = (\omega)((g)\beta)$. Thus $(g)\gamma$ and $(g)\beta$ both act on $\omega \in \Omega$ in the same way, for every $g \in G$ and every $\omega \in \Omega$. Therefore $\beta = \gamma$.

3.11 Note that this is obvious because conjugation by x is an automorphism of G which sends P to P^x. It therefore sends $N_G(P)$ to $N_G(P^x)$. That argument is entirely correct, but it is a little glib. Here is a down-to-earth proof. Suppose that $g \in N_G(P)$, so $h^g \in P$ for every $h \in P$. Now suppose that $k \in P^x$, so $k = t^x$ for some $t \in P$. Thus

$$(g^x)^{-1} k g^x = (g^x)^{-1} t^x g^x = x^{-1} g^{-1} t g x = (t^g)^x.$$

However, $t^g \in P$ so $(t^g)^x \in P^x$. Therefore each element of $N_G(P)^x$ normalizes P^x so $N_G(P)^x \subseteq N_G(P^x)$. Replace x by x^{-1} in this inclusion so $N_G(P)^{x^{-1}} \subseteq N_G(P^{x^{-1}})$. Now replace P by P^x so $N_G(P^x)^{x^{-1}} \subseteq N_G(P^{xx^{-1}}) = N_G(P)$. Conjugate this inclusion by x so $N_G(P^x) \subseteq N_G(P)^x$. We have inclusion in both directions, so $N_G(P)^x = N_G(P^x)$.

3.12 By Sylow's theorem P has $|G : N_G(P)|$ conjugates. Any two distinct conjugates of P intersect in the identity, so the union of conjugates of P has $|G : N_G(P)|(|P|-1)$ non-identity elements. Each one of these elements has p-power order. Conversely, if $g \in G$ has p-power order, then $Q = \langle g \rangle$ is a p-subgroup of G and by Sylow's theorem is a subgroup of some conjugate of P, so g is an element of some conjugate of P. We are done.

3.13 Suppose that $g \in G$, so conjugation by g induces an automorphism of the normal subgroup N. Thus $Q^g \in \text{Syl}_p(N)$. It follows from Sylow's theorem that Q and Q^g are conjugate in N. Thus there is $n \in N$ such that $Q^g = Q^n$ so $Q^{gn^{-1}} = Q$. Therefore $gn^{-1} = x \in N_G(Q)$. Now $g = xn \in N_G(Q)N$. Now $g \in G$ was arbitrary so $G \leq N_G(Q)N \leq G$ and therefore $G = N_G(Q)N$.

3.14 $T \trianglelefteq N_G(T)$ and $P \in \text{Syl}_p(T)$ so the Frattini argument applies. We deduce that $N_G(T) = N_G(P)T = T$ since $N_G(P) \leq T$. We are done.

3.15 Suppose that L is a 1-dimensional subspace of V. For $g \in \mathrm{GL}(V)$ we have $g \cdot L = \{g(l) \mid l \in L\}$. Suppose that L_1, L_2 are distinct 1-dimensional subspaces of V, as are R_1, R_2. Choose non-zero $u_i \in L_i$ and $v_i \in R_i$ for $i = 1, 2$. Now u_1, u_2 are linearly independent, as are v_1, v_2 by the distinctness of the subspaces. Extend u_1, u_2 to a basis u_1, u_2, u_3 of V. Extend v_1, v_2 to a basis v_1, v_2, v_3 of V. There is a (unique) $x \in \mathrm{GL}(V)$ such that $x(u_j) = v_j$ for $j = 1, 2, 3$. Now $x \cdot L_i = R_i$ for $i = 1, 2$ so we are done.

3.16 Let A_1, A_2 be 2-dimensional subspaces of V which intersect in the trivial subspace $\{0\}$. Let B_1, B_2 be 2-dimensional subspaces of V which intersect in a 1-dimensional subspace. Now $A_1 + A_2 = V$ is 4-dimensional, and $B_1 + B_2$ is $2 + 2 - 1 = 3$-dimensional. If (for contradiction) there is $y \in \mathrm{GL}(V)$ such that $y \cdot A_i = B_i$ for $i = 1, 2$, then y has image contained in $B_1 + B_2$ and so is not surjective. Therefore y is singular which is absurd since every $y \in \mathrm{GL}(V)$ is non-singular.

3.17 Suppose that $x_1, \ldots, x_{n-2}$ are distinct elements of the set $\{1, 2, \ldots, n\}$ with the two missing elements being x_{n-1}, x_n. Suppose also that $y_1, \ldots, y_{n-2}$ are distinct elements of $\{1, 2, \ldots, n\}$ with the two missing elements being y_{n-1}, y_n. There is $\sigma \in S_n$ such that $x_i \cdot \sigma = y_i$ for all i in the range $1 \leq i \leq n$. If $\sigma \in S_n$, then we are done. If not, $\sigma' = \sigma(y_{n-1}, y_n) \in A_n$ and $x_i \cdot \sigma' = y_i$ for all i in the range $1 \leq i \leq n - 2$ and we are done.

Chapter 4

4.1 There are q^{n^2} such matrices, since each of the n^2 entries can take any one of q values.

4.2 The first row must be linearly independent, so must not be the zero vector. There are $q^n - 1$ such rows. The second row must be linearly independent of the first row. There are $q^n - q$ such rows for any given first row. Proceeding in this way we see that

$$|\mathrm{GL}(n, q)| = \prod_{i=0}^{n-1} (q^n - q^i).$$

Thus

$$|\mathrm{GL}(3, 2) = (8 - 1)(8 - 2)(8 - 4) = 168.$$

4.3 The first isomorphism theorem tells us that

$$|\mathrm{SL}(n, q)| = |\mathrm{GL}(n, q)|/(q - 1) = (q - 1)^{-1} \prod_{i=0}^{n-1} (q^n - q^i).$$

4.4 $|Z| = t$, so

$$|\mathrm{PSL}(n, q)| = |\mathrm{SL}(n, q)|/t.$$

When $q = 7$ we have $t = 2$ so

$$|\mathrm{PSL}(2, 7)| = (7^2 - 1)(7^2 - 7)/(7 - 1)2 = 168.$$

Chapter 5

5.1 $xyz = zxy[xy, z]$ but also $xyz = xzy[y, z] = zx[x, z]y[y, z] = zxy[x, z][x, z, y][y, z]$ so $[xy, z] = [x, z][x, z, y][y, z]$.

5.2

$$
\begin{aligned}
xyz &= x(yz) = (yz)x[x, yz] = y(zx)[x, yz] \\
&= (zx)y[y, zx][x, yz] = z(xy)[y, zx][x, yz] \\
&= (xy)z[z, xy][y, zx][x, yz] = xyz[z, xy][y, zx][x, yz]
\end{aligned}
$$

so $[z, xy][y, zx][x, yz] = 1$.

5.3 We give a direct proof. Put $r = x^{-1}y^{-1}xz^{-1}x^{-1}$, $s = y^{-1}z^{-1}yx^{-1}y^{-1}$ and $t = z^{-1}x^{-1}zy^{-1}z^{-1}$. Now

$$
[x, y^{-1}, z]^y = (x^{-1}y^{-1}xz^{-1}x^{-1})(yxy^{-1}zy) = rs^{-1}.
$$

Similarly $[y, z^{-1}, x]^z = st^{-1}$ and $[z, x^{-1}, y]^x = tr^{-1}$ so

$$
[x, y^{-1}, z]^y [y, z^{-1}, x]^z [z, x^{-1}, y]^x = rs^{-1}st^{-1}tr^{-1} = 1
$$

and we are done.

5.4 $[h, g] = h^{-1}h^g$. Thus $h^{-1}h^g \in H$ if and only if $h^g \in H$. Thus $H \trianglelefteq G$ if and only if $[h, g] \in H$ for every $h \in H$ and for every $g \in G$.

5.5 In S_n commutators must be even permutations because a permutation and its inverse are either both even or both odd. Thus the commutators must be in A_n and in the case $n = 3$ they are in $\langle(1, 2, 3)\rangle$. Now at least one commutator must be non-trivial since S_3 is non-abelian, and since $(1, 2, 3)$ and $(1, 3, 2)$ are conjugate, they are both commutators. The identity element is always a commutator since id = [id, id]. Of course, you could calculate the answer, but you don't have to do so!

5.6 If H is abelian and $x, y \in G$, then

$$
\begin{aligned}
([x, y])\varphi &= (x^{-1}y^{-1}xy)\varphi = (x)\varphi^{-1} \cdot (y)\varphi^{-1}(x) \cdot \varphi(y) \cdot \varphi \\
&= [(x)\varphi, (y)\varphi] = 1.
\end{aligned}
$$

The final equality is because H is abelian. Thus $C \subseteq \operatorname{Ker} \varphi$. On the other hand, suppose that $C \subseteq \operatorname{Ker} \varphi$. Thus for all $x, y \in G$ we have $1 = ([x, y])\varphi = [(x)\varphi, (y)\varphi]$. However, φ is an epimorphism so $[h, k] = 1$ for all $h, k \in H$ and so H is abelian.

5.7 We may assume that P is not the trivial group, so $Z(P) \neq 1$. Let $Q = P/Z(P)$, a finite p-group which is smaller than P. Select (by inductive hypothesis) a central series for Q. Using the correspondence principle each term of this series is $H/Z(P)$ for a unique subgroup H of P with $Z(P) \leq H \leq P$. The correspondence principle preserves inclusion and normality so we obtain a normal series for P. The correspondence principle respects centralizers, and therefore centres and therefore centrality. Thus we have a central series for P.

5.8 If G is abelian the result is clear. Induct on the nilpotency class. Apply the natural map $G \to G/Z(G)$ to a central series of minimal length for G we obtain a shorter central series for $G/Z(G)$. We have a subnormal series through $HZ(G)/Z(G)$ in $G/Z(G)$ by induction on the class. By the Correspondence Principle we have a subnormal series for G through $HZ(G)$. Now $1 \trianglelefteq H \trianglelefteq HZ(G)$ so we are done.

5.9 M is strictly contained in its normalizer so $M \trianglelefteq G$. Now maximality forces G/M to be a simple group. If G/M is abelian we are done. We induct on the nilpotency class of G. If $Z(G) \leq M$, then it follows that $G/Z(G)$ has smaller class than G, and $M/Z(G)$ is a maximal subgroup of $G/Z(G)$. By induction $G/Z(G)/M/Z(G)$ is abelian and the memorable form of the third isomorphism theorem takes us home. If $Z(G) \nleq M$ then $Z(G)M = G$ by maximality, so $M \trianglelefteq G$ and $G/M = Z(G)M/M \simeq Z(G)/(Z(G) \cap M)$ is abelian and we are done.

5.10 S_1 is the trivial group and has no composition factors. $S_2 \simeq C_2$ is simple and so has a single composition factor C_2. S_3 has a cyclic normal subgroup of index 2 and so has composition factors C_3 and C_2.

5.11 Each generator $[h_1, \ldots, h_k]$ of $\gamma_k(H)$ is in $\gamma_k(G)$ so $\gamma_k(H) \leq \gamma_k(G)$.

5.12 Choose any generator $[g_1, \ldots, g_k]$ of $\gamma_k(G)$. Now

$$([g_1, \ldots, g_k])\theta = [(g_1)\theta, \ldots, (g_k)\theta]$$

is an element of $\gamma_k(H)$. Each element of $\gamma_k(G)$ is a word on the given generators and their inverses, so θ will map this word to a word on elements of $\gamma_k(H)$.

5.13 $\gamma_k(G/N)$ is generated by all elements of the form

$$[Ng_1, Ng_2, \ldots, Ng_k] = N[g_1, g_2, \ldots, g_k].$$

This is the image of $\gamma_k(G)$ under the natural epimorphism $G \to G/N$ and so is $\gamma_k(G)N/N$. We are done.

5.14 This follows immediately from part (i) of Proposition 5.39 using induction.

5.15 $[h, k] = h^{-1}h^k = (k^{-1})^h k \in H \cap K$. Normality is clear since $[h, k]^g = [h^g, k^g] \in [H, K]$ for every $h \in H, k \in K$ and $g \in G$.

5.16 Suppose that $G = H_1 \geq H_1 \geq H_2 \geq \cdots$ with all H_i normal in G and all factors abelian. Then $\delta_i(G) \leq H_i$ by induction on i. In particular $\delta_i(G) \leq G_i$. Now suppose that $G_j = \delta_j(G)$. Now $\delta_j(G)/\delta_{j+1}(G)$ is abelian since the commutators of generators of $\delta_j(G)$ are in $\delta_{j+1}(G)$. Thus the derived group of G_j (which is G_{j+1}) is a subgroup of $\delta_{j+1}(G)$. Thus $G_{j+1} = \delta_{j+1}(G)$ and we are done.

5.17 N must be abelian. If it had any proper non-trivial characteristic subgroup, then this would violate minimality. Therefore H has only one Sylow p-subgroup and so N is a p-group. The elements of N of order p form a non-trivial characteristic subgroup of N, so N has exponent p. The structure theorem for finite abelian groups now applies.

5.18 Let the penultimate term of the derived series of G be A, so A is an abelian normal subgroup of G. Now if $A \leq M$ we finish by looking at G/A and inducting on the derived length. Otherwise $MA = G$ and $|MA : M| = |A : A \cap M|$ has prime power order by the solution to Question 2.

Chapter 6

6.1 Certainly $[a] = [a^2] = [a^3] = \cdots$ so $|M| \leq 2$. The only issue is whether or not $[1] = [a]$. We can show that $|M| = 2$ if we can find a monoid with two elements where the equation is satisfied.

$$\left\{ \begin{pmatrix} 1 & 0 \\ 0 & 1 \end{pmatrix}, \begin{pmatrix} 1 & 0 \\ 0 & 0 \end{pmatrix} \right\}$$

under matrix multiplication will do nicely. Here $0, 1 \in \mathbb{Z}$.

6.2 The relation $\sim'$ suggested is translation invariant and an equivalence relation. Moreover, if $l = r$ is any equation in R, then $l \sim' r$. Thus $\sim_R \subseteq \sim'$. However, if $u \sim' v$ then $u \sim_R v$ so $\sim' \subseteq \sim_R$.

6.3 The constructive version of the equivalence relation outlined in the previous question shows 1 and a are each only equivalent to themselves, so that $1, a$ and a^2 are pairwise inequivalent. Alternatively you can exhibit a monoid of size 3 satisfying the equation. Let a be the 3×3 matrix with integer entries as shown, and I, a, a^2 is such a monoid.

$$a = \begin{pmatrix} 0 & 1 & 0 \\ 0 & 0 & 1 \\ 0 & 0 & 1 \end{pmatrix}.$$

The multiplication table is

	1	a	a^2
1	1	a	a^2
a	a	a^2	a^2
a^2	a^2	a^2	a^2

6.4 Knuth–Bendix yields a five rule complete rewriting system:
  (i) $abc \to a$
  (ii) $bca \to b$
  (iii) $aa \to ab$
  (iv) $bbc \to b$
  (v) $ba \to bb$.

Chapter 8

8.1 Let T be the transitive closure of the symmetric closure of the reflexive closure of R. Let E be the equivalence closure of R. Certainly $T \subset E$. The interesting problem is to show that T is an equivalence relation, because that will force $T = E$. Suppose that $x \in X$. Now (x, x) is in the reflexive closure of R so $(x, x) \in T$. Next suppose that $(x, y) \in T$. Therefore there is a finite sequence of pairs (x_i, x_{i+1}) for $0 \le i < n$ with $x_0 = x$, $x_n = y$ and $(x_i, x_{i+1}) \in S$, the symmetric closure of the reflexive closure of R. Since S is symmetric we have $(x_{i+1}, x_i) \in S$ for every i and so $(y, x) \in T$, the transitive closure of S. Finally T is a transitive relation because it is a transitive closure. Thus T is an equivalence relation.

8.2 One problem is this. Suppose that $(a, b), (a, c) \in R$ where $b \ne c$, then there is no map (or graph of a map) containing both (a, b) and (a, c).

8.3 Let $X = \{1\}$ and suppose R is empty. The empty relation is both transitive and symmetric, but it is not reflexive.

8.4 $|\Omega \times \Omega| = n^2$ so the number of relations on Ω is size of the power set of a set of size n^2. This is 2^{n^2}. Thus there are 512 relations on the set $\{1, 2, 3\}$.

8.5 at, Clements, first, Frank, in, January, noon, on, Square, the, thousand, Trafalgar, two, was, year.

8.6 year, was, two, Trafalgar, thousand, the, Square, on, noon, January, in, Frank, first, Clements, at.

8.7 at, in, on, the, two, was, noon, year, Frank, first, Square, January, Clements, thousand, Trafalgar.

8.8 Suppose that x, y are both minimal elements of S. Since $<$ is linearly ordered we have both $x \le y$ and $y \le x$. Thus $x = y$.

Index

m-cycle, 91
p-groups, 116

abelian, 2
abelianization, 162
action, 103
– transitive, 120
Adian S.I., 41
alternating group, 99
aperiodic, 152
automorphism, 59
axiom of choice, 237
axioms of a field, 219
axioms of a group, 1
axioms of a monoid, 195

block, 123
Brauer–Fowler theorem, 137
Burnside, 41, 109
Burnside's $p^\alpha q^\beta$ theorem, 135
butterfly lemma, 69

Cartesian product, 73
Cauchy's theorem, 113
Cayley table, 12
Cayley's theorem, 107
central element, 48
central series, 168
centralizer, 110
centre, 48
characteristic of a field, 223
characteristic subgroup, 60
Church–Rosser Theorem, 211
classification of finite simple groups, 137
Clements, F., 234

commutative diagram, 63
commutator, 72
commutator calculus, 159
commutator quotient, 162
commutator subgroup, 161
complement, 76
composition factors, 167
composition of maps, 2
composition series, 167
confluence, 211
conjugacy, 46
conjugacy class, 47, 110
conjugacy classes of S_n, 99
core, 116
core of a subgroup, 116
Correspondence Principle, 55
coset representatives, 21, 22
cosets, 19
critical pair, 214
cycle, 91, 92
cycle shape, 98
cyclic group, 32

degree, 89
derived group, 161
derived subgroup, 155
dihedral, 38
direct product, 71

embedding, 64
endomorphism, 51
epimorphism, 51
equivalence class, 230
equivalence relation, 230
equivalent refinements, 166